LECTURES ON
LINEAR LOGIC

CSLI
Lecture Notes
No. 29

LECTURES ON LINEAR LOGIC

A. S. Troelstra

CSLI CENTER FOR THE
STUDY OF LANGUAGE
AND INFORMATION

CSLI was founded early in 1983 by researchers from Stanford University, SRI International, and Xerox PARC to further research and development of integrated theories of language, information, and computation. CSLI headquarters and the publication offices are located at the Stanford site.

CSLI/SRI International   **CSLI/Stanford**   **CSLI/Xerox PARC**
333 Ravenswood Avenue   Ventura Hall   3333 Coyote Hill Road
Menlo Park, CA 94025   Stanford, CA 94305   Palo Alto, CA 94304

Printed in the United States

99 98 97 96 95 94 93 92   5 4 3 2 1

Library of Congress Cataloging-in-Publication Data

Troelstra, A. S. (Anne Sjerp)
    Lectures on linear logic / A. S. Troelstra.
      p.   cm. -- (CSLI lecture notes ; no. 29)
    Includes bibliographical references and index.
    ISBN 0-937073-78-4 -- ISBN 0-937073-77-6 (pbk.)
    1. Logic, Symbolic and mathematical. I. Title. II. Title: Linear logic.
III. Series.
QA9.T76 1991
511.3--dc20   91-38902
                                                 CIP

Preface

The present lecture notes originated in a course on linear logic at the University of Amsterdam, spring semester 1991. A condensed presentation was given at the University of Bern, Switzerland in June 1991. As an appendix to the body of the text, D. Roorda contributed the last chapter. The notes aim at giving an all-around introduction. Some very recent work had to be left out, sometimes with regret, e.g bounded linear logic, most of the work on decidability problems, Girard's work on the "geometry of interaction," etc. Certain syntactical aspects on the other hand have gotten a bit of extra attention, since they can play an important role if one wants to familiarize oneself with linear logic. Of the first seven chapters dealing with syntax, Chapters 4, 5, 7 are not needed in what follows. If these notes are well received, and the subject continues to flourish, then perhaps I will some day find the courage to prepare an updated and expanded version.

Notwithstanding the fact that already many papers have been using J.-Y. Girard's notation for the operators of linear logic, I have ventured, with some trepidation, to introduce some changes in the notation, so as to obtain a better harmony with the notations of lattice theory and the relevant parts of category theory. Most of my changes will not lead to confusion, since I have used different symbols instead of giving a new meaning to symbols already in use for linear

logic. There is one exception: I have interchanged the role of Girard's
0 and ⊥.

Prerequisites. These lectures are intended for readers with some so-
phistication in logic, such as graduate students in logic or theoretical
computer science, or an experienced researcher new to linear logic.
Some familiarity with classical and possibly intuitionistic predicate
logic is necessary, and it is certainly helpful to have seen Gentzen-
style sequent calculi such as LK, LJ and cut elimination for these
systems (see, e.g., the treatment of sequent calculi in Kleene 1952).
As for category theory, we shall assume familiarity with the material
of the following chapters and sections of MacLane 1971: I, II 1–4;
III 1–5; IV 1–2, 4, 6; V 1–5; VI 1; and optional is VII 1–3, 7 since
the relevant notions are explained in the notes. Good preliminary
reading is Girard, Lafont & Taylor 1988.

Acknowledgments. I am indebted to H. Geuvers, R. Hindley, B. Ja-
cobs, J. van Oosten, D. Roorda, R. Stärk, and especially H. Schellinx
for discussions, criticisms, providing proofs, and help in finding the
relevant literature.

Author's address: Faculteit Wiskunde en Informatica
                   Universiteit van Amsterdam
                   Plantage Muidergracht 24
                   1018TV AMSTERDAM (NL)
                   e-mail: anne@fwi.uva.nl

Contents

List of Tables

1 Introduction

1.1. Linear logic, introduced in Girard 1987, is interesting from a purely logical point of view, and potentially of considerable interest for computer science. Linear logic may be viewed as an example of a "resource-conscious" logic, where the formulas represent types of resource, and resources cannot be used ad libitum. That is to say, asserting a sequent $A, A \Rightarrow B$ means something like: we use two data (resources) of type A to obtain one datum of type B. A truth on the other hand is something which can be used freely, as often as we like. In Gentzen-style sequential formalisms "resource-consciousness" shows itself by the absence of (some of) the so-called structural rules. Hence resource-conscious logics are in a Gentzen-style sequent formulation also "substructural logics". In linear logic the possibility of using resources of a certain type arbitrarily often is represented by a special logical operator called storage. In this respect linear logic differs from other substructural logics.

Linear logic is not the only substructural logic; other, historically earlier examples are relevance logic, categorial logic and BCK-logic. Some comments on these related enterprises, with a few references, are given in Chapter 2. A *warning* for those who have some preliminary acquaintance with papers on linear logic: at certain points we *deviate from Girard's original notation*, which is used in many papers (for a motivation of the choice of notation, see Section 2.7).

1

1.2. The well known logical formalism of intuitionistic propositional logic permits two quite distinct, but not unrelated intuitive interpretations: a logical one and a type-theoretic one. From the logical viewpoint, the formulas stand for propositions or statements, and the rules tell us how to derive correct conclusions from correct premises.

In the type-theoretic interpretation, the formulas stand for certain sets (e.g., $A \wedge B$ corresponds to the cartesian product of the sets (corresponding to) A and B); a proof of A from premises $B_1, \ldots, B_n$ encodes a method of finding an element of set A from elements of sets $B_1, \ldots, B_n$. The realization that intuitionistic propositional logic permits both interpretations is expressed by the slogan "formulas-as-types", or "propositions-as-types".

A proof of A from B, seen as a method (i.e., a function) for constructing elements of A from elements of B, permits the use of any finite number of elements of B in the construction of an element of A, in keeping with the fact that we permit in the logical formalism any finite number of uses of a given assumption. Obviously, from the computational viewpoint, we might be interested in a more careful (less wasteful) bookkeeping, where we keep track of the number of times a datum of a given type is used. As we shall see, this corresponds in Gentzen sequent calculi to the deletion of the so-called "structural" rules of weakening and contraction.

1.3. In the discussion below, as well as later on in these notes, we shall often need the concept of a *multiset*. Intuitively, a multiset is a set with (finite) multiplicities; there may be finitely many "copies" of a single element. As a formal definition we use

DEFINITION. A *multiset* over A is a mapping $f : A \longrightarrow \mathbb{N}$, where $f(a) = n$ means that a occurs with multiplicity n. If $f(a) = 0$, a is not an element of f (that is to say, a occurs with multiplicity 0). $\square$

1.4. In order to introduce the basic ideas, we shall consider the "baby example" of the pure theory of conjunction ($\wedge$). The pure theory starts from a countable supply of propositional variables; formulas are constructed with $\wedge$.

The logical theory of conjunction can be formulated in various ways. Below we discuss two distinct types of formulation: sequent calculi with left and right introduction rules, and natural deduction with introduction and elimination rules.

1.5. The intuitionistic calculus of conjunction

Let us use greek upper case Γ, Γ', Γ'', ..., Δ, Δ', ... for finite sequences of formulas, A, B, C, ... for arbitrary formulas. A *sequent* is an expression $\Gamma \Rightarrow A$. The calculus is given by specifying axioms and rules.

(i) Axioms are all sequents $A \Rightarrow A$.

(ii) Structural rules:

$$\frac{\Gamma, A, B, \Gamma' \Rightarrow C}{\Gamma, B, A, \Gamma' \Rightarrow C} \text{ (exchange)}$$

$$\frac{\Gamma \Rightarrow C}{\Gamma, A \Rightarrow C} \text{ (weakening or thinning)}$$

$$\frac{\Gamma, A, A \Rightarrow C}{\Gamma, A \Rightarrow C} \text{ (contraction)}$$

(iii) Logical rules :

$$\frac{\Gamma \Rightarrow A_0 \qquad \Gamma \Rightarrow A_1}{\Gamma \Rightarrow A_0 \wedge A_1} \text{ (R}\wedge \text{ or right-}\wedge\text{-introduction)}$$

$$\frac{\Gamma, A_i \Rightarrow C}{\Gamma, A_0 \wedge A_1 \Rightarrow C} \text{ (L}\wedge \text{ or left-}\wedge\text{-introduction, } i \in \{0, 1\})$$

Observe that if we read the Γ, Γ', ... Δ, Δ', ... as *multisets* instead of sequences (i.e., sequences modulo the ordering) the rule of exchange need not be stated; and if we interpret the Γ, Γ', ... as *finite sets*, we can also drop the rule of contraction. Finally, if we generalize the axioms to

$$\Gamma, A \Rightarrow A$$

and read the Γ, Γ', ... as finite sets, we can drop all three structural rules and obtain an equivalent system, that is to say, the same sequents are derivable as in the original system.

These versions of the calculus are equivalent as to derivable sequents, but not combinatorially: passing from sequences to multisets, for example, means erasing the applications of exchange and thus identifying distinct proof trees (the differences are of a fairly trivial nature however).

Another equivalent version of the calculus is obtained by replacing the logical rules by e.g.,

$$\text{L}\wedge' \ \frac{\Gamma, A_0, A_1 \Rightarrow B}{\Gamma, A_0 \wedge A_1 \Rightarrow B} \qquad \text{R}\wedge' \ \frac{\Gamma_0 \Rightarrow A_0 \qquad \Gamma_1 \Rightarrow A_1}{\Gamma_0, \Gamma_1 \Rightarrow A_0 \wedge A_1}$$

The proof of equivalence of this new version with the old one *essentially* uses the structural rules, in particular weakening and contraction:

$$\frac{\dfrac{\dfrac{\Gamma, A_0, A_1 \Rightarrow B}{\Gamma, A_0 \wedge A_1, A_1 \Rightarrow B}}{\Gamma, A_0 \wedge A_1, A_0 \wedge A_1 \Rightarrow B}}{\Gamma, A_0 \wedge A_1 \Rightarrow B}$$

For the converse we use weakening.

Similarly, an application of R$\wedge$ is transformed into an application of R$\wedge'$ followed by contractions (and exchanges).

CONVENTION. We adopt, unless stated otherwise, the convention that $\Gamma, \Gamma', \ldots, \Delta, \Delta', \ldots$ are finite multisets, so that the exchange rule need not appear explicitly any more. In sequent notation the empty multiset is often denoted by an empty place (e.g., $\Gamma \Rightarrow$); otherwise we use Λ. $\square$

1.6.   Note that in the absence of contraction and weakening, the choice of rules for conjunction leads to distinct connectives (conjunction analogues). Thus if we combine R$\wedge'$ with L$\wedge'$, we have a "context-free" version of conjunction (henceforth to be denoted by $\star$, and called "tensor" or "times"), i.e.,

$$\text{R}\star \; \frac{\Gamma_0 \Rightarrow A_0 \qquad \Gamma_1 \Rightarrow A_1}{\Gamma_0, \Gamma_1 \Rightarrow A_0 \star A_1} \qquad \text{L}\star \; \frac{\Gamma, A_0, A_1 \Rightarrow B}{\Gamma, A_0 \star A_1 \Rightarrow B}$$

If on the other hand we choose the pair R$\wedge$, L$\wedge$ we obtain a context-sensitive conjunction analogue (henceforth denoted by $\sqcap$, called "and")

$$\text{R}\sqcap \; \frac{\Gamma \Rightarrow A_0 \qquad \Gamma \Rightarrow A_1}{\Gamma \Rightarrow A_0 \sqcap A_1} \qquad \text{L}\sqcap \; \frac{\Gamma, A_i \Rightarrow B}{\Gamma, A_0 \sqcap A_1 \Rightarrow B} \quad (i \in \{0, 1\})$$

$\sqcap$ is context-sensitive in the sense that in R$\sqcap$ we have a restraint on the contexts in the application of the rule: the same antecedent Γ has to appear in both premises. As we shall see in the next chapter, both connectives $\sqcap$ and $\star$ with the four rules R$\sqcap$, L$\sqcap$, R$\star$, L$\star$ form part of intuitionistic linear logic.

What about connectives obtained by combining R$\wedge$,L$\wedge'$ or R$\wedge'$ and L$\wedge$? We consider this question in the next section.

♠ EXERCISE. Show that for the intuitionistic calculus of conjunction, the version with Γ, Δ sets, and axioms $\Gamma, A \Rightarrow A$ is equivalent to the calculus as described in 1.5.

1.7. Cut elimination and conservativity

We may add to the $\star$-calculus (based on R$\star$, L$\star$) or the $\sqcap$-calculus (based on R$\sqcap$, L$\sqcap$), as described above, a rule

$$\text{Cut} \quad \frac{\Gamma, A \Rightarrow B \qquad \Gamma' \Rightarrow A}{\Gamma, \Gamma' \Rightarrow B}$$

(A is the *cut formula*). This rule can be eliminated by successive transformations of proof trees; the crucial step consists in the replacement of a Cut by one or more Cuts with cut formula of lower complexity. Thus

$$\text{Cut} \quad \frac{\dfrac{\Gamma, A_0 \Rightarrow C}{\Gamma, A_0 \sqcap A_1 \Rightarrow C} \qquad \dfrac{\Gamma' \Rightarrow A_0 \qquad \Gamma' \Rightarrow A_1}{\Gamma' \Rightarrow A_0 \sqcap A_1}}{\Gamma, \Gamma' \Rightarrow C}$$

is transformed into

$$\frac{\Gamma, A_0 \Rightarrow C \qquad \Gamma' \Rightarrow A_0}{\Gamma, \Gamma' \Rightarrow C} \quad \text{Cut}$$

This also works for the $\star$-calculus. But if we introduce, in the absence of structural rules, a conjunction-analogue "c" with say R$\wedge$ and L$\wedge'$, i.e.,

$$\frac{\Gamma \Rightarrow A_0 \qquad \Gamma \Rightarrow A_1}{\Gamma \Rightarrow A_0 c A_1} \qquad \qquad \frac{\Gamma, A_0, A_1 \Rightarrow C}{\Gamma, A_0 c A_1 \Rightarrow C}$$

this reduction step fails in the *absence* of contraction.

Cut elimination is a desirable property for propositional logics, since it usually entails the subformula property: in a derivation of a sequent occur only subformulas of the formulas in the sequent.

As a result, adding cut to the system with $\star$ and/or $\sqcap$ is *conservative* in the following sense. Let us call a sequent *atomic* if all formulas in $\Gamma \cup \{A\}$ are atomic. It is obvious that with the rules for $\star$ and $\sqcap$ no new atomic sequents become derivable from a given set of atomic sequents $\mathcal{X}$, since each application of a $\star$-rule or $\sqcap$-rule results in the introduction of a *compound* formula.

But it is conceivable that adding the Cut rule would change this, since then a compound formula may appear in the deduction which is removed by applications of Cut. But cut elimination guarantees that this cannot happen; Cut adds nothing to the derivable sequents and a fortiori introduces no new atomic sequents.

On the other hand, adding a connective such as "c" is not harmless: such an addition is not conservative. For example, we can now derive contraction by

$$\frac{\dfrac{A \Rightarrow A \quad A \Rightarrow A}{A \Rightarrow A\,c\,A} \quad \dfrac{\Gamma, A, A \Rightarrow C}{\Gamma, A\,c\,A \Rightarrow C}}{\Gamma, A \Rightarrow C}$$

Similarly, adding c with rules corresponding to $R\wedge'$, $L\wedge$ makes the weakening rule derivable.

Instead of the Cut rule as stated above we might also consider a "context-sensitive" variant (additive Cut rule, see exercise below). But if we think of the logic we are aiming at as a "logic of actions" (use data of types A, B etc. to obtain datum of type B) as indicated in our introductory paragraphs, Cut as stated obviously represents composition of actions, while the variant does not.

♠ EXERCISES.
1. Show that the applications of Cut can be eliminated from the $\star$-calculus and the $\sqcap$-calculus.
2. Consider the following variant Cut_a of Cut:

$$\text{Cut}_a \quad \frac{\Gamma \Rightarrow A \quad \Gamma, A \Rightarrow C}{\Gamma \Rightarrow C}$$

(the additive Cut rule) and show that the $\star$-calculus is not closed under this rule, e.g., by showing that $A \star A \Rightarrow (A \star A) \star (A \star A)$ is derivable with Cut_a, but not without it.

1.8. Natural deduction

A quite different system for $\wedge$-logic is based on natural deduction; this is especially suitable for discussing the formulas-as-types idea. Let us write $\mathcal{D}, \mathcal{D}', \mathcal{D}'', \ldots$ for deductions in this system; we write

$$\begin{array}{c} \mathcal{D} \\ A \end{array}$$

to indicate that $\mathcal{D}$ has conclusion A. Each deduction is a tree with formulas as labels of the nodes; the labels of the top nodes are assumptions. Then the one-point tree with label

$$A$$

is a deduction of A from assumption A. If

$$\begin{array}{c} \mathcal{D}_i \\ A_i \end{array} \quad (i=0,1)$$

are deductions from assumptions Γ_i, then

$$\frac{\begin{array}{cc} \mathcal{D}_0 & \mathcal{D}_1 \\ A_0 & A_1 \end{array}}{A_0 \wedge A_1} \wedge I \; (\wedge\text{-introduction})$$

is a deduction of $A_0 \wedge A_1$ from Γ_0, Γ_1; and if

$$\begin{array}{c} \mathcal{D} \\ A_0 \wedge A_1 \end{array}$$

is a deduction of $A_0 \wedge A_1$ from Γ, then

$$\frac{\begin{array}{c} \mathcal{D} \\ A_0 \wedge A_1 \end{array}}{A_i} \wedge E \; (\wedge\text{-elimination}; \; i \in \{0,1\})$$

is a deduction of A_i from Γ.

Of course, here again we may read the collection of assumptions either as a set or as a multiset. For our baby example this does not make much of a difference. (It is not obvious how to read the collection of assumptions as a *sequence* in a natural way, and we shall disregard this possibility altogether. The distinction between set and multiset becomes much more important combinatorially if we add e.g., implication.)

Proof trees as described above may also be written in sequent notation, i.e., carrying the open assumptions along at each node; then the natural deduction proof trees are generated from *axioms*

$$A \Rightarrow A$$

by means of the *rules*

$$\frac{\Gamma_0 \Rightarrow A_0 \qquad \Gamma_1 \Rightarrow A_1}{\Gamma_0, \Gamma_1 \Rightarrow A_0 \wedge A_1} \quad (\wedge I = R\wedge')$$

$$\frac{\Gamma \Rightarrow A_0 \wedge A_1}{\Gamma \Rightarrow A_i} \quad (\wedge E, i = 0, 1).$$

♠ EXERCISE. Prove the equivalence between the natural deduction calculus for conjunction and the original sequent calculus, in the following sense: $\Gamma \Rightarrow A$ is derivable in the sequent calculus if A is derivable in the natural deduction calculus from Γ' for some set Γ' contained in the multiset Γ.

1.9. Term notation for natural deduction

We may also write deductions as terms, as follows.
(i) With an assumption A we associate a variable of type A

$$x^A : A \text{ or } x^A \vdash A$$

(ii) Rules construct new deduction trees, i.e., terms, from old ones, according to

$$\frac{t_0 : A_0 \qquad t_1 : A_1}{\pi t_0 t_1 : A_0 \wedge A_1} \, (\wedge\mathrm{I})$$

$$\frac{t : A_0 \wedge A_1}{\pi_i t : A_i} \, (\wedge\mathrm{E}; \ i \in \{0, 1\})$$

Here π (pairing) and π_0, π_1 (unpairing) are new constants for describing the constructions on deduction trees. Each term obtained in this way completely describes a deduction; the term is in fact nothing but an alternative notation for the proof tree. Strictly speaking, there are distinct π^{A_0, A_1}, $\pi_0^{A_0, A_1}$, $\pi_1^{A_0, A_1}$ for each choice of A_0, A_1, but it is not necessary to show this in the notation, since one easily sees that the types of the various occurrences of π, π_0, π_1 in a term are easily reconstructed provided we know the types of the variables. (The type of π in $\wedge I$ above is $A_0, A_1 \Rightarrow A_0 \wedge A_1$, and of the π_i $A_0 \wedge A_1 \Rightarrow A_i$). If on the other hand we drop the type subscript for variables, we may have to add superscripts to the π, π_i in order to obtain a unique type for each subterm in a given $t : A$.

Note that this notation suggests an interpretation of the collections of open assumptions corresponding neither to the "set" interpretation nor to the "multiset" interpretation. Assigning distinct variables to occurrences of an assumption formula A distinguishes between the occurrences; assigning the same variable to distinct occurrences of assumption A may be said to collapse the occurrences into one.

Deduction trees may contain unnecessary detours, namely where an introduction is immediately followed by an elimination, thus

$$\frac{\dfrac{\mathcal{D}_0}{A_0} \qquad \dfrac{\mathcal{D}_1}{A_1}}{\dfrac{A_0 \wedge A_1}{A_i}}$$

The detour may be removed by a contraction to

$$\begin{array}{c} \mathcal{D}_i \\ A_i \end{array}$$

or in term notation

(∗) $\pi_i \pi(t_0, t_1)$ contracts to t_i.

The notation is reminiscent of (and intended to be so) a pairing operator π with decoding (unpairing) operators π_0, π_1. This analogy is elaborated in the formulas-as-types idea.

1.10. Formulas-as-types

The "formulas-as-types" idea (f.a.t.i) is the idea that a formula may be identified with (is characterized by) the set of its proofs. As is familiar from the informal (Brouwer-Heyting-Kolmogorov-) interpretation, the (constructive) proof of a conjunction $A_0 \wedge A_1$ is given by a pair of proofs, one for A_0 and one for A_1. So on the f.a.t.i. we think of $A_0 \wedge A_1$ as a cartesian product, and the π, π_i can then be taken to be the pairing operators and the first and second projections respectively. (∗) in the preceding section then simply expresses that the π_i act as projections.

On the other hand, (∗) does not fully express the idea of $A_0 \wedge A_1$ as a cartesian product, but only $A_0 \times A_1 \subset A_0 \wedge A_1$, i.e., that the cartesian product is *contained* in $A_0 \wedge A_1$; in order to express that every element of $A_0 \wedge A_1$ is a pair we also need

SP $\pi(\pi_0 t, \pi_1 t) = t$

(*surjectivity of pairing*), besides

PROJ $\pi_i \pi(t_0, t_1) = t_i.$

SP does not so readily suggest itself as a contraction on proof trees; it corresponds to

$$\frac{\dfrac{\mathcal{D}}{A_0 \wedge A_1} \qquad \dfrac{\mathcal{D}}{A_0 \wedge A_1}}{A_0 \wedge A_1} \quad \text{contracts to} \quad \frac{\mathcal{D}}{A_0 \wedge A_1}$$

By addition of PROJ and SP we have completed the turn from a logical view of the $\wedge$-calculus to a type-theoretic one.

The type-theoretic approach suggests models different from the model suggested by the logical approach. Thus, where logically we tend to think of $\bot$, falsehood, as having an empty set of proofs, on the type-theoretic view $\bot$ may very well be represented by some non-empty set. Natural deduction calculus and f.a.t.i. for intuitionistic linear logic are discussed in Chapters 6 and 13.

1.11. From set-theoretic types to a categorical approach

The type-theoretic way of looking at things suggests a further generalization: the categorical approach. The cartesian product may be generalized to the categorical product. The details, for the theory of conjunction, are as follows. We read any sequent $\Gamma \Rightarrow A$ as $\bigwedge \Gamma \Rightarrow A$, where $\bigwedge(A_1, \ldots, A_n) := (\ldots ((A_1 \wedge A_2) \wedge A_3) \ldots)$; so all sequents are interpreted as sequents with a single formula in the antecedent. Let us call such sequents *1-sequents*. The following set of axioms and rules generates a system equivalent w.r.t. the set of derivable 1-sequents:

$$A \Rightarrow A \qquad \qquad \frac{A \Rightarrow B \qquad B \Rightarrow C}{A \Rightarrow C}$$

$$A_0 \wedge A_1 \Rightarrow A_i \qquad \frac{B \Rightarrow A_0 \qquad B \Rightarrow A_1}{B \Rightarrow A_0 \wedge A_1}$$

We may, just as for natural deduction, introduce a system of terms as a notation for deductions in this system; we also replace $\Rightarrow$ by the categorical arrow $\longrightarrow$.

$$\mathrm{id}_A : A \longrightarrow A \qquad \frac{t : A \longrightarrow B \qquad s : B \longrightarrow C}{s \circ t : A \longrightarrow C}$$

$$\pi_i : A_0 \wedge A_1 \longrightarrow A_i \qquad \frac{t_0 : B \longrightarrow A_0 \qquad t_1 : B \longrightarrow A_1}{\langle t_0, t_1 \rangle : B \longrightarrow A_0 \wedge A_1}$$

We may think of the resulting terms as describing arrows in a category with products (the reason behind our replacement of $\Rightarrow$ by $\longrightarrow$), where the objects correspond to formulas. However, this interpretation introduces some obvious identifications between the various terms denoting arrows.

Suppose $t : A \longrightarrow B$, $t' : B \longrightarrow C$, $t'' : C \longrightarrow D$. Then we must satisfy

(i) *category axioms*

$$t \circ \mathrm{id}_A = t, \qquad \mathrm{id}_B \circ t = t,$$

$$t'' \circ (t' \circ t) = (t'' \circ t') \circ t.$$

(ii) *product axioms*

$$\pi_i \circ \langle t_0, t_1 \rangle = t_i,$$

$$\langle \pi_0 \circ t, \pi_1 \circ t \rangle = t.$$

Instead of the axioms under (ii) we may take

$$\pi_i \circ \langle t_0, t_1 \rangle = t_i, \quad \langle \pi_0, \pi_1 \rangle = \mathrm{id}_{A_0 \wedge A_1},$$

$$\langle t_0, t_1 \rangle \circ t_2 = \langle t_0 \circ t_2, t_1 \circ t_2 \rangle.$$

Let $\mathcal{X}$ be a directed graph, with arrows (= edges) x, x', x'', ... A free category $F(\mathcal{X})$ with binary products may be constructed over $\mathcal{X}$ by forming all possible expressions from the arrows of $\mathcal{X}$ with help of $\circ$, $\langle\ \rangle$, π_0, π_1, and taking equivalence classes w.r.t. the least equivalence relation generated by the identities (i) and (ii).

If we want to have categories with all finite products, it suffices to postulate a terminal object $\top$ with unique arrows from A to $\top$ for any object A, so

$$\top_A : A \longrightarrow \top$$

$$t = \top_A \text{ for any } t : A \longrightarrow \top$$

As we shall see, $\sqcap$ in 1.6 corresponds to categorical products, but $\star$ corresponds to a tensor product.

Intuitionistic propositional logic corresponds in the same way to cartesian closed categories with finite coproducts, and intuitionistic linear logic, to be described in the next chapter, corresponds to symmetric monoidal closed categories with finite products and coproducts.

The category-theoretic point of view, which may be seen as a generalization of the type-theoretic viewpoint, suggests questions of the following kind:

(a) Can we decide whether an arrow $t : A \longrightarrow B$ exists for given A, B?

(b) Can we decide the equality between arrows $t, t' : A \longrightarrow B$?

(c) Suppose the graph $\mathcal{X}$ is a category. Is the embedding functor from $\mathcal{X}$ into $F(\mathcal{X})$ full and faithful?

The categorical approach to linear logic is treated in Chapters 9, 11, 12.

♠ EXERCISE. Show that the two sets of equations for categorical products are equivalent.

1.12. The computational aspects of intuitionistic and linear logic

Already the natural-deduction rules for conjunction logic reveal an (admittedly trivial) computational aspect, via the term notation for deductions and the formulas-as-types ideas. Deduction trees may be normalized by removing detours (i.e., introductions immediately followed by eliminations), that is to say terms of the form $\pi_i\pi(t_0, t_1)$ are replaced by t_i. So the normalization of deductions in the $\wedge$ calculus corresponds to a calculus of pairing and unpairing.

If, instead of intuitionistic conjunction logic, we take intuitionistic implication logic, that is to say in addition to axioms $x^A : A$ we have rules

$$\to\text{I} \ \frac{t[x^A] : B}{\lambda^A x.t : A \to B} \qquad\qquad \to\text{E} \ \frac{t : A \to B \qquad t' : A}{tt' : B}$$

for introducing and eliminating implications, then removal of a detour, namely replacement of

$$\frac{\dfrac{t : B}{\lambda x^A.t : A \to B} \qquad t' : A}{(\lambda x^A.t)t' : B}$$

by $t[x^A/t'] : B$, is in fact a β-conversion in a typed lambda-calculus. Each deduction term can be "evaluated", i.e., transformed into normal form by successive β-conversions. Hence intuitionistic propositional logic corresponds computationally to typed lambda-calculus with *additional* operators (such as pairing and unpairing corresponding to conjunction).

In quite the same manner, namely via a suitable term calculus with evaluation rules corresponding to normalization steps on deductions we can give a computational interpretation to an "intuitionistic" fragment of linear logic (Chapters 13–15).

In the proofnets of Girard, introduced in Chapter 17 we encounter the computational side of linear logic in a different form (Chapter 18).

The connection between linear logic and the theory of Petri nets, briefly discussed below, reveals a link with parallel computation.

1.13. Petri nets and the theory of tensor $\star$

In these notes we do not intend to go very deeply into the connections between linear logic and Petri nets; we limit ourselves to a brief description of the connection between the theory of Petri nets and the pure theory of tensor. More about the connection between Petri nets and linear logic in Brown 1990, Brown and Gurr 1990, Engberg and Winskel 1990, Martí-Oliet and Meseguer 1989b, 1991.

DEFINITION. A Petri net is a triple $\mathcal{N} \equiv (S, T, F)$, where S is a set of *places*, T a set of *transitions*, $S \cap T = \emptyset$, and F is a multiset over $(S \times T) \dot\cup (T \times S)$, where $\dot\cup$ denotes disjoint union. $\square$

Graphically we can represent a Petri net by drawing the places as circles, the transitions as squares, and an arrow from place A to transition t [from transition t to place A] labeled with $n \in \mathbb{N} \setminus 0$ if $(A, t) \in F$ [$(t, A) \in F$] with multiplicity n ($n = 1$ usually omitted).

Definition. A *marking* of a Petri net $\mathcal{N} \equiv (S, T, F)$ is a multiset over S.

1.14. Example. A marking can be indicated in a graphical representation of a Petri net by inscribing the multiplicities in the circles. We show a Petri net $\mathcal{N} \equiv (S, T, X)$ with $S = \{A, B, C, D, E, F\}$, $T = \{t1, t2\}$, $X(A, t1) = 1$, $X(B, t1) = 3, \ldots, X(t2, F) = 1$, marked with $M(A) = 2, M(B) = 5, \ldots$. Graphically this becomes

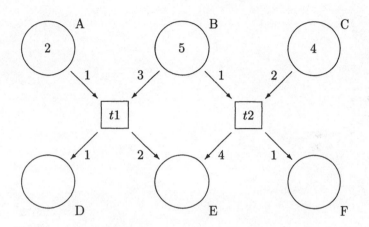

A "firing" of a multiset of transitions transforms the given marking into another one; firing a single transition t subtracts (adds) n from the mark m in place X if there is an arrow with label n from X to t (t to X). Firing a multiset of transitions gives the result of firing the transitions of the multiset in some order.

Firing $t1$ and $t2$ in this example results in a marking $M'(A) = 1, M'(B) = 1, M'(C) = 2, M'(D) = 1, M'(E) = 6, M'(F) = 1$.

1.15.   In the theory of $\star$ as given in 1.6, we can prove $\star$ to be associative and commutative. Hence we may systematically identify $X \star (Y \star Z)$ with $(X \star Y) \star Z$, and $X \star Y$ with $Y \star X$, and accordingly write simply $A^1 \star B^3$, or even $A^1 B^3$, for $((A \star B) \star B) \star B)$, etc.

It is now quite simple to translate a given Petri net in a tensor theory with axioms corresponding to the transitions. Thus the Petri net of our example corresponds to a theory with the general axiom

$$\text{id} : X \Rightarrow X$$

(for any string X constructed from the places $A,B,C,\ldots$ and $\star$), two specific axioms

$$t1 \; : \; A^1 \star B^3 \Rightarrow D^1 \star E^2$$
$$t2 \; : \; B^1 \star C^2 \Rightarrow E^4 \star F.$$

and two general rules

$$\frac{t : A \Rightarrow B \qquad t' : B \Rightarrow C}{t' \circ t : A \Rightarrow C} \qquad\qquad \frac{t : A \Rightarrow B \qquad t' : C \Rightarrow D}{t \star t' : A \star C \Rightarrow B \star D}$$

(the first rule corresponds to sequential composition, the second to parallel composition).

The result of firing the multiset $\{t1, t2\}$ is now found by the following deduction:

$$\frac{\dfrac{t1 : AB^3 \Rightarrow DE^2 \qquad t2 : BC^2 \Rightarrow E^4 F}{t1 \star t2 : AB^4C^2 \Rightarrow DE^6 F} \qquad id : ABC^2 \Rightarrow ABC^2}{id \star (t1 \star t2) : A^2 B^5 C^4 \Rightarrow ABC^2 DE^6 F}$$

Conversely, tensor theories of the type just indicated can be interpreted straightforwardly as describing the behaviour of finite Petri nets. More precisely, let $P_0, \ldots, P_n$ be propositional variables, and let W, W', W_i, W_i' be tensor products of these variables. If we take $W = W'$ to mean that W and W' are equal or differ only in the grouping and order of the factors (which means that symmetry and associativity of tensor is built in), we may treat them as multisets.

Now let $\mathbf{T}$ be a theory based on the axiom id, sequential and parallel composition as above, and on finitely many axioms

$$\text{Ax}_i \qquad\qquad W_i \Rightarrow W_i' \quad (0 \leq i < m)$$

which may be used arbitrarily often in a deduction. $\mathbf{T}$ codes the behaviour of a finite Petri net with places $P_0, \ldots, P_n$ and transitions t_i corresponding to the Ax_i as in our example above, and the derivability of $W \Rightarrow W'$ means that the marking coded by W' can be reached from the marking coded by W in this Petri net. ($P_0^{s(0)} \ldots P_n^{s(n)}$ codes the marking M given by $M(P_i) = s(i)$).

The "reachability problem" for finite Petri nets is the question of deciding whether a marking M' can be reached from a given marking M; the correspondence outlined above shows that this is equivalent to the deducibility problem for sequents in finitely axiomatized $\star$-theories. This problem is known to be decidable (see e.g., Mayr & Meyer 1981 or Clote 1986); full propositional linear logic is undecidable (see Chapter 20).

2 Sequent calculus for linear logic

2.1. Chapters 2–5 study classical and intuitionistic linear logic as logics (i.e., as collections of derivable sequents and formulas) from a syntactical point of view. In the present chapter we introduce sequent calculi in the style of Gentzen's LK and LJ, that is to say with left and right introduction rules.

Our notation for the logical operators adopted here differs from Girard's notation; see our remarks on notation in 2.7.

The adjectives "intuitionistic" and "classical" as applied to linear logic are primarily motivated by formal analogy to Gentzen's LJ and LK respectively. In fact, classical linear logic **CLL** is also highly constructive: as we shall see in Chapter 5 intuitionistic logic can be faithfully embedded into classical linear logic. Even if the subtheory of intuitionistic linear logic **ILL** lacks some of the elegant symmetry of **CLL**, the system may nevertheless claim some independent interest, having interesting models which cannot obviously be expanded into models of **CLL**.

2.2. The sequent calculus for classical linear logic

Capitals $A, B, \ldots$ denote formulas, $\Gamma, \Gamma', \ldots, \Delta, \Delta', \ldots$ finite multisets of formulas. The empty multiset is usually indicated by a blank, or by a Λ if necessary to avoid confusion. A sequent with antecedent Γ and succedent Δ is written as $\Gamma \Rightarrow \Delta$. Prime (= atomic) formulas are as usual for first-order predicate logic; compound formulas are constructed with the help of

$$\sqcap, \ \sqcup, \ \top, \ \bot, \ \star, \ \mathbf{1}, \ +, \ \mathbf{0}, \ \multimap, \ \sim, \ !, \ ?, \ \forall, \ \exists.$$

These operators fall into four groups.

(i) the *multiplicative* or *context-free* operators: $\star$ (*times, tensor* or *tensor product*), $+$ (*par*, i.e., "parallel or"), $\multimap$ (*linear implication*), $\mathbf{0}$ (*zero*), $\mathbf{1}$ (*unit*). For context-free operators the side formulas of the premises of any rule application are simply added in the conclusion.

(ii) The *additive* or *contextual* operators: $\sqcap$ (*and* or *conjunction*), $\sqcup$ (*or* or *disjunction*), $\top$ (the *true*) and $\bot$ (the *false*).

In rules for additive constants with several premises, the side formulas in each of the premises coincide with the side formulas of the conclusion.

Both multiplicative and additive constants are characterized by explicit rules: the constant appears in the rule schemata only once, namely as main operator in a formula of the conclusion, and each rule involves only a single operator. This is different in the third group below.

(iii) The *exponentials* or *modalities*: ! (*storage* or *ofcourse*) and ? (*whynot, consumption* or *costorage*). These operators are superficially similar to the modal operators $\Box$, $\Diamond$ in the usual modal logics. It is to be noted that the rules R! and L? in the system below are not explicit, since ! and ? occur in the side formulas.

The role of ! and ? is to introduce weakening and contraction in a controlled way for individual formulas: ! for the antecedent, ? for the succedent.

By inspection of the rules we see immediately that if we add *weakening* and *contraction*

$$\text{LW} \ \frac{\Gamma \Rightarrow \Delta}{\Gamma, A \Rightarrow \Delta} \qquad \text{RW} \ \frac{\Gamma \Rightarrow \Delta}{\Gamma \Rightarrow \Delta, A}$$

$$\text{LC} \ \frac{\Gamma, A, A \Rightarrow \Delta}{\Gamma, A \Rightarrow \Delta} \qquad \text{RC} \ \frac{\Gamma \Rightarrow A, A, \Delta}{\Gamma \Rightarrow A, \Delta}$$

then we may interpret ! and ? by the identity, i.e., $!A := ?A := A$ validates all the rules.

(iv) The first-order quantifiers $\forall$ and $\exists$ behave pretty much as in ordinary predicate logic.

It is suggestive to think of the formulas as data types (linear logic as a "book-keeping method for data"). Some examples of the reading of the connectives on this interpretation: a datum of type $A \sqcap B$ is a datum which can be used *once* to extract either a datum of type

A or a datum of type B; a datum of type $A \star B$ is a pair of data; a datum of type $A \sqcup B$ contains a datum of type A or a datum of type B (we do not know in advance which). A datum of type $A \multimap B$ is a method of transforming a *single* datum of type A into a datum of type B. $!A$ indicates that we may extract as many data of type A as we like, i.e., a datum of type $!A$ is a finite collection of data of type A. See also Chapter 19.

2.3. CONVENTION. Unary operators and quantifiers bind stronger than binary operators; $\star$, $+$, $\sqcup$, $\sqcap$ bind stronger than $\multimap$. $\square$

We are now ready to present the rules and axioms of **CLL** in Table 1. For comparison we also give the sequent calculus for classical logic **CL** (i.e., Gentzen's LK) in Table 2.

2.4. Some comments on the system

(i) There is a lot of symmetry in the system **CLL**, and it is therefore not surprising that we have a form of de Morgan duality. The following pairs of operators (F, F^*) are *dual* to each other (that is to say $\sim F(P, P', \ldots)$ is equivalent to $F^*(\sim P, \sim P', \ldots)$, and F^{**} is identical with F):

$$(\sim, \sim),\ (\star, +),\ (\mathbf{1}, \mathbf{0}),\ (\sqcap, \sqcup),\ (\top, \bot),\ (\forall, \exists),\ (!, ?).$$

In addition, $\multimap$ is definable from $+$ and $\sim$:

$$A \multimap B := \sim A + B$$

(ii) The three rules W!, L!, C! may be replaced by

$$!\text{Ax} \qquad\qquad !A \Rightarrow \mathbf{1} \sqcap A \sqcap (!A \star !A).$$

For example, W! is derivable from !Ax by

$$
\frac{
\dfrac{\dfrac{\Gamma \Rightarrow \Delta}{\mathbf{1}, \Gamma \Rightarrow \Delta}}{\mathbf{1} \sqcap (A \sqcap (!A \star !A)), \Gamma \Rightarrow \Delta} \qquad !A \Rightarrow \mathbf{1} \sqcap (A \sqcap (!A \star !A))
}{
!A, \Gamma \Rightarrow \Delta
} \text{Cut}
$$

The rest is left as an exercise.

TABLE 1

The sequent calculus for classical linear logic **CLL**

Logical axiom and Cut rule:

$$\text{Ax} \quad A \Rightarrow A \qquad \text{Cut} \frac{\Gamma \Rightarrow A, \Delta \quad \Gamma', A \Rightarrow \Delta'}{\Gamma, \Gamma' \Rightarrow \Delta, \Delta'}$$

Rules for the propositional constants:

$$\text{L}\sim \frac{\Gamma \Rightarrow A, \Delta}{\Gamma, \sim A \Rightarrow \Delta} \qquad\qquad \text{R}\sim \frac{\Gamma, A \Rightarrow \Delta}{\Gamma \Rightarrow \sim A, \Delta}$$

$$\text{L}\sqcap \frac{\Gamma, A \Rightarrow \Delta \quad \Gamma, B \Rightarrow \Delta}{\Gamma, A \sqcap B \Rightarrow \Delta \quad \Gamma, A \sqcap B \Rightarrow \Delta} \qquad \text{R}\sqcap \frac{\Gamma \Rightarrow A, \Delta \quad \Gamma \Rightarrow B, \Delta}{\Gamma \Rightarrow A \sqcap B, \Delta}$$

$$\text{L}\star \frac{\Gamma, A, B \Rightarrow \Delta}{\Gamma, A \star B \Rightarrow \Delta} \qquad \text{R}\star \frac{\Gamma \Rightarrow A, \Delta \quad \Gamma' \Rightarrow B, \Delta'}{\Gamma, \Gamma' \Rightarrow A \star B, \Delta, \Delta'}$$

$$\text{L}\sqcup \frac{\Gamma, A \Rightarrow \Delta \quad \Gamma, B \Rightarrow \Delta}{\Gamma, A \sqcup B \Rightarrow \Delta} \qquad \text{R}\sqcup \frac{\Gamma \Rightarrow A, \Delta}{\Gamma \Rightarrow A \sqcup B, \Delta} \quad \frac{\Gamma \Rightarrow B, \Delta}{\Gamma \Rightarrow A \sqcup B, \Delta}$$

$$\text{L}+ \frac{\Gamma, A \Rightarrow \Delta \quad \Gamma', B \Rightarrow \Delta'}{\Gamma, \Gamma', A + B \Rightarrow \Delta, \Delta'} \qquad \text{R}+ \frac{\Gamma \Rightarrow A, B, \Delta}{\Gamma \Rightarrow A + B, \Delta}$$

$$\text{L}\multimap \frac{\Gamma \Rightarrow A, \Delta \quad \Gamma', B \Rightarrow \Delta'}{\Gamma, \Gamma', A \multimap B \Rightarrow \Delta, \Delta'} \qquad \text{R}\multimap \frac{\Gamma, A \Rightarrow B, \Delta}{\Gamma \Rightarrow A \multimap B, \Delta}$$

$$\text{L1} \frac{\Gamma \Rightarrow \Delta}{\Gamma, 1 \Rightarrow \Delta} \qquad\qquad \text{R1} \quad \Rightarrow 1$$

$$\text{(no L}\top) \qquad\qquad \text{R}\top \quad \Gamma \Rightarrow \top, \Delta$$

$$\text{L0} \quad 0 \Rightarrow \qquad\qquad \text{R0} \frac{\Gamma \Rightarrow \Delta}{\Gamma \Rightarrow \Delta, 0}$$

$$\text{L}\bot \quad \Gamma, \bot \Rightarrow \Delta \qquad\qquad \text{(no R}\bot)$$

Rules for the quantifiers (y not free in Γ, Δ):

$$\text{L}\forall \frac{\Gamma, A[x/t] \Rightarrow \Delta}{\Gamma, \forall x\, A \Rightarrow \Delta} \qquad \text{R}\forall \frac{\Gamma \Rightarrow A[x/y], \Delta}{\Gamma \Rightarrow \forall x\, A, \Delta}$$

$$\text{L}\exists \frac{\Gamma, A[x/y] \Rightarrow \Delta}{\Gamma, \exists x\, A \Rightarrow \Delta} \qquad \text{R}\exists \frac{\Gamma \Rightarrow A[x/t], \Delta}{\Gamma \Rightarrow \exists x\, A, \Delta}$$

Rules for the exponentials:

$$\text{W!} \frac{\Gamma \Rightarrow \Delta}{\Gamma, !A \Rightarrow \Delta} \qquad \text{L!} \frac{\Gamma, A \Rightarrow \Delta}{\Gamma, !A \Rightarrow \Delta} \qquad \text{R!} \frac{!\Gamma \Rightarrow A, ?\Delta}{!\Gamma \Rightarrow !A, ?\Delta} \qquad \text{C!} \frac{\Gamma, !A, !A \Rightarrow \Delta}{\Gamma, !A \Rightarrow \Delta}$$

$$\text{W?} \frac{\Gamma \Rightarrow \Delta}{\Gamma \Rightarrow ?A, \Delta} \qquad \text{L?} \frac{!\Gamma, A \Rightarrow ?\Delta}{!\Gamma, ?A \Rightarrow ?\Delta} \qquad \text{R?} \frac{\Gamma \Rightarrow A, \Delta}{\Gamma \Rightarrow ?A, \Delta} \qquad \text{C?} \frac{\Gamma \Rightarrow ?A, ?A, \Delta}{\Gamma \Rightarrow ?A, \Delta}$$

TABLE 2
The sequent calculus for classical logic **CL**

Logical axiom and Cut rule:

$$\text{Ax } \ A \Rightarrow A \qquad\qquad \text{Cut } \ \frac{\Gamma \Rightarrow A, \Delta \qquad \Gamma', A \Rightarrow \Delta'}{\Gamma, \Gamma' \Rightarrow \Delta, \Delta'}$$

Rules for the propositional constants:

$$\text{L}\sim \frac{\Gamma \Rightarrow A, \Delta}{\Gamma, \sim A \Rightarrow \Delta} \qquad\qquad \text{R}\sim \frac{\Gamma, A \Rightarrow \Delta}{\Gamma \Rightarrow \sim A, \Delta}$$

$$\text{L}\wedge \frac{\Gamma, A \Rightarrow \Delta \qquad \Gamma, B \Rightarrow \Delta}{\Gamma, A \wedge B \Rightarrow \Delta \quad \Gamma, A \wedge B \Rightarrow \Delta} \qquad \text{R}\wedge \frac{\Gamma \Rightarrow A, \Delta \qquad \Gamma \Rightarrow B, \Delta}{\Gamma \Rightarrow A \wedge B, \Delta}$$

$$\text{L}\vee \frac{\Gamma, A \Rightarrow \Delta \qquad \Gamma, B \Rightarrow \Delta}{\Gamma, A \vee B \Rightarrow \Delta} \qquad \text{R}\vee \frac{\Gamma \Rightarrow A, \Delta}{\Gamma \Rightarrow A \vee B, \Delta} \qquad \frac{\Gamma \Rightarrow B, \Delta}{\Gamma \Rightarrow A \vee B, \Delta}$$

$$\text{L}\rightarrow \frac{\Gamma \Rightarrow A, \Delta \qquad \Gamma, B \Rightarrow \Delta}{\Gamma, A \rightarrow B \Rightarrow \Delta} \qquad \text{R}\rightarrow \frac{\Gamma, A \Rightarrow B, \Delta}{\Gamma \Rightarrow A \rightarrow B, \Delta}$$

Rules for the quantifiers (y not free in Γ, Δ):

$$\text{L}\forall \frac{\Gamma, A[x/t] \Rightarrow \Delta}{\Gamma, \forall x \, A \Rightarrow \Delta} \qquad \text{R}\forall \frac{\Gamma \Rightarrow A[x/y], \Delta}{\Gamma \Rightarrow \forall x \, A, \Delta}$$

$$\text{L}\exists \frac{\Gamma, A[x/y] \Rightarrow \Delta}{\Gamma, \exists x \, A \Rightarrow \Delta} \qquad \text{R}\exists \frac{\Gamma \Rightarrow A[x/t], \Delta}{\Gamma \Rightarrow \exists x \, A, \Delta}$$

Structural rules :

$$\text{LW} \frac{\Gamma \Rightarrow \Delta}{\Gamma, A \Rightarrow \Delta} \qquad \text{LC} \frac{\Gamma, A, A \Rightarrow \Delta}{\Gamma, A \Rightarrow \Delta}$$

$$\text{RW} \frac{\Gamma \Rightarrow \Delta}{\Gamma \Rightarrow A, \Delta} \qquad \text{RC} \frac{\Gamma \Rightarrow A, A, \Delta}{\Gamma \Rightarrow A, \Delta}$$

2.5. The sequent calculus ILL for intuitionistic linear logic

Intuitionistic linear logic is a subsystem of **CLL** of independent interest. Formally, it is obtained from **CLL** in the same way as Gentzen's LJ is obtained from LK (our **CL**), namely by restricting the multisets on the right hand side of the sequents (the succedents) to at most one formula occurrence; in addition, we drop operators and constants with rules essentially involving more than one formula occurrence on the right hand side, that is to say we drop ?, $\sim$, **0** and +. We use **IL** to designate intuitionistic predicate logic (Gentzen's LJ).

Note that in the absence of $\perp$ no sequents with empty succedent are derivable in **ILL**. It is sometimes convenient to consider an version of **ILL** with an extra constant **0**:

DEFINITION. *Intuitionistic linear logic with zero* **ILZ** is the system obtained by adding to **ILL** a constant **0** with an axiom and a rule

$$\text{L0} \quad \mathbf{0} \Rightarrow \qquad\qquad \text{R0}\ \frac{\Gamma \Rightarrow}{\Gamma \Rightarrow \mathbf{0}}$$

□

2.6. REMARKS. These are precisely the axiom and rule for **0** in **CLL** under the "intuitionistic restriction" of at most one formula in the succedent. **ILZ** is in certain cases more suitable for comparison with **CLL** than **ILL**. We shall encounter examples of this situation in Chapter 5.

Alternatively, we may define **ILZ** as an extension of **ILL** with a negation $\sim$ satisfying

$$\text{L}\sim \frac{\Gamma \Rightarrow A}{\Gamma, \sim A \Rightarrow} \qquad\qquad \text{L}\sim \frac{\Gamma, A \Rightarrow}{\Gamma \Rightarrow \sim A}$$

again corresponding to the rules for $\sim$ in **CLL** subject to "the intuitionistic restriction". The versions of **ILZ** are intertranslatable by either defining $\sim A$ as $A \multimap \mathbf{0}$ or $\mathbf{0}$ as ~ 1. The following trivial proposition is to be noted:

PROPOSITION. **ILZ** $\vdash \Gamma \Rightarrow$ iff **ILZ** $\vdash \Gamma \Rightarrow \mathbf{0}$.   □

REMARK. In fact, it is easy to check that in **ILL** with L0 added R0 is a derivable rule, i.e., $\vdash \Gamma \Rightarrow \Lambda$ implies $\vdash \Gamma \Rightarrow \mathbf{0}$. We prefer keeping R0 however, since this gives the proper version for **ILZ** plus additional axioms and guarantees that the proposition also holds for the extension.

2.7. Remarks on the choice of notation

In this version of these lectures on linear logic, we have kept Girard's symbols $\sim$, $\forall$, $\exists$, !, ?, $\top$, $\mathbf{1}$, and replaced &, $\wp$, $\otimes$, $\oplus$ by $\sqcap$, $+$, $\star$, $\sqcup$ respectively, and interchanged $\perp$ and $\mathbf{0}$. Instead of $A^{\perp}$ we write $\sim A$. Thus we have achieved that the pairs $\sqcap$, $\sqcup$; $+$, $\star$; !, ?; $\forall$, $\exists$; $\mathbf{0}$, $\mathbf{1}$; $\perp$, $\top$ are de Morgan duals, where each pair consists of "similar" symbols. With the ordering $\leq$ on the set of equivalence classes $\mathcal{F}$ of formulas modulo linear equivalence, i.e.,

$$[A] = [B] \text{ iff } \vdash \Rightarrow A \multimap B \text{ and } \vdash \Rightarrow B \multimap A$$

and

$$[A] \leq [B] \text{ iff } \vdash \Rightarrow A \multimap B, \text{ or equivalently } \vdash A \Rightarrow B$$

we obtain a lattice with $\top$, $\perp$, $\sqcup$, $\sqcap$ corresponding to top, bottom, join and meet respectively.

We have chosen $\sqcup, \sqcap$ instead of $\vee, \wedge$, since on the one hand the shape of $\sqcup, \sqcap$ is reminiscent of $\vee, \wedge$ (and so $\sqcup$ suggests join and disjunction, and $\sqcap$ suggests meet and conjunction), and on the other hand we do not want to identify $\sqcup$ with disjunction and $\sqcap$ with conjunction in classical or intuitionistic logic, since in linear logic $\vee, \wedge$ have in fact two analogues each (a multiplicative and an additive one). An additional advantage (to me) is that the shape of $\sqcup, \sqcap$ is also reminiscent of $\coprod, \prod$, widely used for categorical coproducts and products respectively. We thought it better to use $\star$ instead of $\otimes$, since $\otimes$ inevitably evokes $\oplus$ of Girard's notation, a source of confusion we wished to avoid.

$(\mathcal{F}, \star, \mathbf{1})$ is a commutative monoid with unit; in classical linear logic, $(\mathcal{F}, +, \mathbf{0})$ is a dual commutative monoid. Since $\star$, $+$ suggest "times" and "plus", it is only natural to use $\mathbf{1}$ and $\mathbf{0}$ for the respective neutral elements.

Girard's choice of notation was motivated differently. Since $\star$ behaves as a tensor product in algebra, he choose $\otimes$; $\sqcup$ was reminiscent of direct sum, hence the choice of $\oplus$. This led to a familiar-looking distributive law:

$$A \otimes (B \oplus C) \text{ is equivalent to } (A \otimes B) \oplus (A \otimes C)$$

which is thus easily memorized. This choice also dictated the use of $\mathbf{1}$ and $\mathbf{0}$ as the respective neutral elements. On the other hand, the fact that $\otimes$ and $\oplus$ are not dual to each other is confusing, as is the fact that not $\top$, $\perp$ but $\top$, $\mathbf{0}$, and not $\mathbf{1}$, $\mathbf{0}$, but $\mathbf{1}$, $\perp$ are duals when one uses Girard's notation.

There is little confusion to be expected from our change of &, $\wp$, $\otimes$, $\oplus$ into $\sqcap$, $+$, $\star$, $\sqcup$, since the symbols are disjoint. On the other hand, the interchange between $\perp$ and $\mathbf{0}$ may cause confusion when using this text next to papers in Girard's notation.

A completely new set of four symbols for the neutral elements here designated by $\top, \perp, \mathbf{1}, \mathbf{0}$ could avoid this confusion, but we were unable to find an alternative which is mnemonically convenient and fits the other symbols as well as the present set. The best alternative seems to be one proposed by K. Došen, namely to use $\mathbf{T}, \mathbf{F}, \mathbf{t}, \mathbf{f}$ for the present $\top, \perp, \mathbf{1}, \mathbf{0}$ respectively. Ordinary "true" splits into additive $\mathbf{T}$ and multiplicative $\mathbf{t}$; "false" similarly splits into $\mathbf{F}$ and $\mathbf{f}$. (One might follow Došen's proposal in discussions with someone accustomed to Girard's notation, so as to avoid confusion.)

Since already quite a number of papers and reports have been written using Girard's original notation, we propose a change of notation with some hesitation, though it seems to us there are some obvious advantages, especially to someone new to the subject. Of course, one can get used to almost any notation provided it is not too cumbersome, and unambiguous!

2.8. Other types of substructural logic

(a) *Relevance logics.* The beginnings of relevance logic date from the early fifties; a good introduction is Dunn 1986. An important motivation for relevance logic stems from the search for a notion of deduction where B is said to follow from A only in case A has actually been used in the deduction of B (this excludes e.g., $A \to (B \to A)$ as a logical theorem). Many systems of relevance logic have been formulated; a representative system is $\mathbf{R}$, due to N.D. Belnap. In Dunn 1986 $\mathbf{R}$ is axiomatized in Hilbert-style by the following axioms and rules:

Implication

$$A \to A,$$
$$(B \to C) \to ((A \to C) \to (A \to C)),$$
$$(A \to (B \to C)) \to (B \to (A \to C)),$$
$$(A \to (A \to B)) \to (A \to B) \text{ (contraction)};$$

Conjunction

$$A \wedge B \to A,$$
$$A \wedge B \to B,$$
$$(A \to B) \wedge (A \to C) \to (A \to B \wedge C),$$

Disjunction

$$A \to A \vee B,$$
$$B \to A \vee B,$$
$$(A \to C) \wedge (B \to C) \to (A \vee B \to C),$$
$$A \wedge (B \vee C) \to (A \wedge B) \vee C \text{ (distribution)},$$

Truth-constants

$$(\mathbf{t} \to A) \leftrightarrow A,$$
$$A \to \mathbf{T},$$

Fusion

$$(A \circ B \to C) \leftrightarrow (A \to (B \to C)),$$

Negation

$$(A \to {\sim}A) \to {\sim}A,$$
$$(A \to {\sim}B) \to (B \to {\sim}A),$$
$${\sim}{\sim}A \to A.$$

Rules

$$A, A \to B \Rightarrow B \text{ (modus ponens)},$$
$$A, B \Rightarrow A \wedge B \text{ (adjunction)}.$$

In this system the logical operators and constants $\to, \wedge, \vee, \mathbf{t}, \mathbf{T}, \circ, \sim$ correspond to $\multimap, \sqcap, \sqcup, \mathbf{1}, \top, \star, \sim$ in linear logic. Negation may be defined as $A \to \mathbf{f}$, where $\mathbf{f}$ is a new constant for falsehood, which has to satisfy the axiom schema $((A \to \mathbf{f}) \to \mathbf{f}) \to A$. Conversely, from $\sim$ we can define $\mathbf{f}$ as $\sim\mathbf{t}$. $\mathbf{f}$ corresponds to $\mathbf{0}$ in linear logic.

Modulo this correspondence of the logical symbols, $\mathbf{R}$ has an equivalent axiomatization by the axioms of contraction and distribution, and the axioms (1)-(13) of the Hilbert-type axiomatization H-$\mathbf{CLL}_0$ of $\mathbf{CLL}_0$ in Chapter 7, and the rules of modus ponens and adjunction. Alternatively, $\mathbf{R}$ might be axiomatized by the sequentcalculus for the $\{\multimap, \sqcap, \sqcup, \star, \mathbf{1}, \top, \sim\}$-fragment of $\mathbf{CLL}_0$, with the contraction rule and the distribution axiom added.

Distribution especially often poses technical problems, and makes comparison with work in linear logic difficult.

(b) *Categorial logics.* The prototype of a categorial logic is the syntactic calculus of Lambek 1958 (although the idea of categorial grammar arose before then). The motivation for this type of logic is the use of logic as a grammar for natural languages. The formulas of the logic are syntactic categories; basic categories are e.g., e (*entity*) and t (*truth-value*). Thus in "Anne sobs", Anne may be given type e, the whole sentence should be of type t, and so "sobs" is a function from entities to truth-values, of type $e \backslash t$. So $e \backslash t$ stands for the category of expressions which combined with an expression of type e on the *left-hand side* yield an expression of type t. What about "the child sobs"? "The child" is not simply an entity, it is of a higher abstraction level; the natural thing to do is to give it type $t/(e\backslash t)$, the type of expressions which yield expressions of type t when combined on the *right* with an expression of type $e\backslash t$. However, "Anne", when combined with an expression of type $e\backslash t$ on the right also yields an expression of type t. We therefore expect a law $e \Rightarrow t/(e\backslash t)$: if an expression is of type e, we also want it to have type $t/(e\backslash t)$. Note that we have two functional types from A to B, namely $B\backslash A$ (argument of type B appears on the *left*) and A/B (argument of type B appears on the *right*).

The original Lambek calculus **L** concerns just these two functional types, and has as axioms and rules ($\Gamma, \Gamma', \Delta, \Delta', \dots$ finite sequences of formulas)

$$\text{Ax} \quad A \Rightarrow A$$

$$\text{L}\backslash \ \frac{\Gamma \Rightarrow A \quad \Delta, B, \Delta' \Rightarrow C}{\Delta, \Gamma, A\backslash B, \Delta' \Rightarrow C} \qquad \text{R}\backslash \ \frac{A, \Gamma \Rightarrow B}{\Gamma \Rightarrow A\backslash B}$$

$$\text{L}/ \ \frac{\Gamma \Rightarrow A \quad \Delta, B, \Delta' \Rightarrow C}{\Delta, B/A, \Gamma, \Delta' \Rightarrow C} \qquad \text{R}/ \ \frac{\Gamma, A \Rightarrow B}{\Gamma \Rightarrow B/A}$$

there are no structural rules, not even exchange. The law $e \Rightarrow t/(e\backslash t)$ is now derivable:

$$\frac{\dfrac{e \Rightarrow e \quad t \Rightarrow t}{e, e\backslash t \Rightarrow t}}{e \Rightarrow t/(e\backslash t)}$$

Clearly, adding exchange makes $A\backslash B$ and A/B equivalent:

$$\frac{\dfrac{\dfrac{A \Rightarrow A \quad B \Rightarrow B}{A, A\backslash B \Rightarrow B}}{A\backslash B, A \Rightarrow B}}{A\backslash B \Rightarrow B/A}$$

and the resulting calculus is equivalent to the $\multimap$-fragment of intuitionistic linear logic. In other words, in **L** $\multimap$ is split into $\backslash$ and $/$.

Of course one may add further connectives: already in Lambek 1958 a "tensor" is mentioned, with rules

$$\text{R} \star \ \frac{\Gamma \Rightarrow A \quad \Delta \Rightarrow B}{\Gamma, \Delta \Rightarrow A \star B} \qquad \text{L} \star \ \frac{\Gamma, A, B, \Delta \Rightarrow C}{\Gamma, A \star B, \Delta \Rightarrow C}$$

Ultimately, by adding enough constants, on obtains something like a non-commutative version of $\textbf{ILL}_0$.

Categorial grammar is not the only motivation for studying this kind of formalism. Just as the $\star$, $\multimap$-fragment of **ILL** is connected with the theory of symmetric closed monoidal categories (cf. Chapter 9), so the theory of $\star$, $/$ corresponds to the theory of monoidal closed categories.

One may get an impression of the manifold aspects of categorial grammars from the papers in Oehrle et. al. 1988 (especially for the connections with natural languages) and van Benthem 1991.

(c) *Logics without contraction.* Examples are e.g., Grishin 1974, Ketonen & Weyhrauch 1984, Ono & Komori 1985. Logics without contraction are often called "BCK-logics", since the implicational fragment of these logics may be axiomatized by axiom schemes corresponding to the principal-type schemes of the combinators B,C,K in a Curry-style assignment system (cf. 6.10). The calculi L_{BCC} and L_{BCK} studied in Ono & Komori 1985 may be obtained by taking the sequent calculus for the $\{\multimap, \sqcup, \sqcap, \star, \perp\}$-fragment of **ILL**, and adding weakening (for L_{BCC}), or weakening and exchange (for L_{BCK}). Grishin 1974 considers a system which is essentially CLL_0 with weakening added.

It is therefore not surprising that these systems are in many respects very close to linear logic — the addition of weakening to linear logic has less dramatic consequences than the addition of contraction has.

The three groups of logics listed above do not exhaust the area of logics with technical similarities to linear logic. Just to mention an example, Casari (1987, 1989) introduced "comparative logic", in an attempt to axiomatize the notion "x is less A than y" (A a property), with a primitive $\leq$ ($\alpha \leq \beta$ expresses that the degree of truth of α is less than the degree of truth of β) which behaves as an implication. Because of the quite different intuition behind these systems, the qualification "resource-conscious logic" seems less appropriate in this case.

3 Some elementary syntactic results

The present section is devoted to some elementary syntactic results concerning **CLL** and **ILL**: some derivable sequents, replacement of equivalents, the one-sided calculus for **CLL**, cut elimination, etc. As sources we used Avron 1988, Roorda 1989,1991, and Schellinx 1990.

3.1. NOTATION. We write $\mathbf{CLL}_0$, $\mathbf{ILL}_0$, $\mathbf{ILZ}_0$ for the purely propositional subsystems of **CLL**, **ILL** without quantifiers and exponentials. Similarly we use $\mathbf{CLL}_q$, $\mathbf{ILL}_q$, $\mathbf{ILZ}_q$ if the quantifiers are included but the exponentials excluded, and $\mathbf{CLL}_e$, $\mathbf{ILL}_e$, $\mathbf{ILZ}_e$ if the quantifiers are excluded but the exponentials included. (The subscripts $_q$ and $_e$ serve to remind us of "quantifier" and "exponential" respectively.) □ In the literature one uses sometimes **MALL** for $\mathbf{CLL}_0$ (from "multiplicative-additive linear logic").

3.2. NOTATION.

$$A \multimap\!\!\circ B := (A \multimap B) \sqcap (B \multimap A) \quad \text{(linear equivalence)}$$

$$A \Leftrightarrow B := A \Rightarrow B \text{ and } B \Rightarrow A. \quad \square$$

N.B. $\vdash\; \Rightarrow A \multimap\!\!\circ B$ iff $\vdash A \Leftrightarrow B$.

3.3. PROPOSITION. *(Provable sequents)*

(i) In **CLL** *and* **ILL***: If* $\vdash\; \Rightarrow A_1 \multimap (A_2 \multimap \ldots (A_n \multimap B)\ldots)$, *then* $\vdash A_1, \ldots, A_n \Rightarrow B$.

26

(ii) In $\mathbf{ILL}_0$ *the following sequents are derivable*

(1) $\Rightarrow A \multimap A$

(2) $A \multimap (B \multimap C) \Rightarrow B \multimap (A \multimap C)$ *(antecedent permutation)*

(3) $A \multimap B,\ B \multimap C \Rightarrow A \multimap C$ *(transitivity)*

(4) $A \sqcap B \Rightarrow A,\ A \sqcap B \Rightarrow B$

(5) $(A \multimap B) \sqcap (A \multimap C) \Leftrightarrow A \multimap B \sqcap C$

(6) $\Rightarrow A \multimap (B \multimap A \star B)$

(7) $A \multimap (B \multimap C) \Leftrightarrow A \star B \multimap C$

(8) $A \Rightarrow A \sqcup B,\ B \Rightarrow A \sqcup B$

(9) $(A \multimap C) \sqcap (B \multimap C) \Leftrightarrow A \sqcup B \multimap C$

(10) $\Rightarrow \mathbf{1},\ \mathbf{1} \Rightarrow (A \multimap A)$,

(11) $\Rightarrow \perp \multimap A, \Rightarrow A \multimap \top$

(iii) In $\mathbf{CLL}_0$ *we can prove:*

(12) $\Rightarrow\, \sim\!\mathbf{0},\ \Rightarrow A \star \sim\!A \multimap \mathbf{0}$,

(13) $A \Leftrightarrow\, \sim\!\sim\!A,\ \Rightarrow A \circ\!\!\!\circ \sim\!\sim\!A$

(14) $A \multimap \sim\!B \Leftrightarrow B \multimap \sim\!A$

(15) $A + B \Leftrightarrow\, \sim\!A \multimap B$

(16) $\sim\!A \multimap B \Leftrightarrow\, \sim(\sim\!A \star \sim\!B)$

(17) $A \star (B \sqcup C) \Leftrightarrow (A \star B) \sqcup (A \star C)$.

(iv) In $\mathbf{CLL}$ *we can prove de Morgan duality (i.e.,* $F(P,\ldots) \Leftrightarrow \sim\!F^*(\sim\!P,\ldots))$ *for all pairs mentioned in 2.4.*

PROOF. (i) We give the proof for $n = 2$. Assume $\Rightarrow A_1 \multimap (A_2 \multimap B)$. With L$\multimap$ the axioms $C \Rightarrow C$, $D \Rightarrow D$ yield $C, C \multimap D \Rightarrow D$. Hence $A_1, A_1 \multimap (A_2 \multimap B) \Rightarrow A_2 \multimap B$; the assumption with the Cut rule yields $A_1 \Rightarrow A_2 \multimap B$; applying Cut again with $A_2 \multimap B, A_2 \Rightarrow B$ we find $A_1, A_2 \Rightarrow B$.

(ii)–(iv). Completely straightforward. $\square$

3.4. PROPOSITION

(i) In **ILL$_0$** *the operators* $\sqcap$, $\sqcup$, $\star$ *are symmetric and associative ; in* **CLL$_0$** *the same holds for* +; *thus e.g.,*

$$A \sqcap B \Leftrightarrow B \sqcap A, \quad A \sqcap (B \sqcap C) \Leftrightarrow (A \sqcap B) \sqcap C.$$

(ii) $\top$, $\bot$, **1**, **0** *are neutral elements for* $\sqcap$, $\sqcup$, $\star$, + *respectively, i.e.,*

$$A \sqcap \top \Leftrightarrow A, \quad A \sqcup \bot \Leftrightarrow A, \quad A \star \mathbf{1} \Leftrightarrow A, \quad A + \mathbf{0} \Leftrightarrow A.$$

For + *this is provable in* **CLL$_0$**, *for the others in* **ILL$_0$**.
(iii) In **ILL$_0$** $\sqcap$ *and* $\sqcup$ *are idempotent, i.e.,*

$$A \sqcup A \Leftrightarrow A, \quad A \sqcap A \Leftrightarrow A.$$

PROOF. Straightforward. $\square$

CONVENTION. We may drop parentheses in repeated uses of $\sqcap$, $\sqcup$, $\star$, +; this is justified by (i) of the proposition. $\square$
The next proposition collects some facts involving the exponentials:

3.5. PROPOSITION

(i) In **ILL** *we can prove*

(1) $!A \Rightarrow \mathbf{1} \sqcap A \sqcap (!A \star !A)$

(2) $!(A \sqcap B) \Leftrightarrow !A \star !B$

(3) $!(A \sqcap B) \Rightarrow !A \sqcap !B, \quad !(!A \sqcap !B) \Leftrightarrow !(A \sqcap B)$

(4) $!\forall x\, A \Rightarrow \forall x !A, \quad !\forall x !A \Leftrightarrow !\forall x\, A$

(5) $\mathbf{1} \Leftrightarrow !\top$

(ii) In **CLL** *we can prove*

(6) $\mathbf{0} \Leftrightarrow\ \sim !\top.$

(7) $?(A \sqcup B) \Leftrightarrow ?A + ?B$

(8) $?A \sqcup ?B \Rightarrow ?(A \sqcup B), \quad ?(?A \sqcup ?B) \Leftrightarrow ?(A \sqcup B)$

(9) $\exists x ?A \Rightarrow ?\exists x\, A, \quad ?\exists x\, A \Leftrightarrow ?\exists x ?A$

PROOF. Mostly straightforward. As an example we prove (2):

$$
\dfrac{\dfrac{A \Rightarrow A}{A \sqcap B \Rightarrow A}}{\dfrac{!(A \sqcap B) \Rightarrow A}{!(A \sqcap B) \Rightarrow !A}} \qquad
\dfrac{\dfrac{B \Rightarrow B}{A \sqcap B \Rightarrow B}}{\dfrac{!(A \sqcap B) \Rightarrow B}{!(A \sqcap B) \Rightarrow !B}}
$$

$$
\dfrac{!(A \sqcap B), !(A \sqcap B) \Rightarrow !A \star !B}{!(A \sqcap B) \Rightarrow !A \star !B}
$$

$$
\dfrac{\dfrac{\dfrac{A \Rightarrow A}{!A \Rightarrow A} \qquad \dfrac{B \Rightarrow B}{!B \Rightarrow B}}{\dfrac{!A, !B \Rightarrow A \qquad !A, !B \Rightarrow B}{!A, !B \Rightarrow (A \sqcap B)}}}{\dfrac{!A, !B \Rightarrow !(A \sqcap B)}{\dfrac{!A \star !B \Rightarrow !(A \sqcap B)}{!A \star !B \Rightarrow !(A \sqcap B)}}} \qquad \square
$$

3.6. Inversion of rules

It is a familiar fact from proof theory of classical logic that certain rules are invertible, e.g., $\Gamma \Rightarrow A \sqcap B, \Delta$ iff $\Gamma \Rightarrow A, \Delta$ and $\Gamma \Rightarrow B, \Delta$. A more refined version ("the inversion lemma") yields, for a suitable notion of depth or length of a deduction

$$
\text{If } \vdash_n \Gamma \Rightarrow A \sqcap B, \Delta \text{ then } \vdash_n \Gamma \Rightarrow A, \Delta \text{ and } \vdash_n \Gamma \Rightarrow B, \Delta
$$

where $\vdash_n$ expresses derivability by means of a deduction of depth or length at most n. For linear logic we can formulate a similar result; more useful is the following version based on a special complexity measure.

DEFINITION. For derivations $\mathcal{D}$ in **CLL** or **ILL** we define a *measure* $d(\mathcal{D})$ as follows:

  (i) $d(\mathcal{D}) = 0$ if $\mathcal{D}$ is an axiom;

 (ii) $d(\mathcal{D}) = d(\mathcal{D}') + 1$ if $\mathcal{D}$ is obtained from $\mathcal{D}'$ by one of the rules L⊓, R⊔, L∼, L∀, R∃, or an exponential rule;

(iii) $d(\mathcal{D}) = \max(d(\mathcal{D}'), d(\mathcal{D}'')) + 1$ if $\mathcal{D}$ is obtained from $\mathcal{D}', \mathcal{D}''$ by one of the rules R⋆, L+, L⊸, Cut;

 (iv) $d(\mathcal{D}) = d(\mathcal{D}')$ if $\mathcal{D}$ is obtained from $\mathcal{D}'$ by one of the rules R+, R⊸, R∼, R0, R∀, L⋆, L1, L∃;

  (v) $d(\mathcal{D}) = \max(d(\mathcal{D}'), d(\mathcal{D}''))$ if $\mathcal{D}$ is obtained from $\mathcal{D}', \mathcal{D}''$ by one of the rules R⊓, L⊔.

We write $\vdash_n \Gamma \Rightarrow \Delta$ if $\Gamma \Rightarrow \Delta$ is derivable from atomic instances of Ax by a derivation $\mathcal{D}$ with $d(\mathcal{D}) \leq n$. $\square$

3.7. Proposition. *(Inversion lemma) In* **CLL**, *and as far as applicable also in* **ILL**,

(i) $\vdash_n \Gamma \Rightarrow A \multimap B, \Delta$ iff $\vdash_n \Gamma, A \Rightarrow B, \Delta$;

(ii) $\vdash_n \Gamma, A \star B \Rightarrow \Delta$ iff $\vdash_n \Gamma, A, B \Rightarrow \Delta$;

(iii) $\vdash_n \Gamma \Rightarrow A + B, \Delta$ iff $\vdash_n \Gamma \Rightarrow A, B, \Delta$;

(iv) $\vdash_n \Gamma \Rightarrow {\sim}A, \Delta$ iff $\vdash_n \Gamma, A \Rightarrow \Delta$;

(v) $\vdash_n \Gamma \Rightarrow \mathbf{0}, \Delta$ iff $\vdash_n \Gamma \Rightarrow \Delta$;

(vi) $\vdash_n \Gamma, \mathbf{1} \Rightarrow \Delta$ iff $\vdash_n \Gamma \Rightarrow \Delta$;

(vii) $\vdash_n \Gamma \Rightarrow A \sqcap B, \Delta$ iff $\vdash_n \Gamma \Rightarrow A, \Delta$ and $\vdash_n \Gamma \Rightarrow B, \Delta$;

(viii) $\vdash_n \Gamma, A \sqcup B \Rightarrow \Delta$ iff $\vdash_n \Gamma, A \Rightarrow \Delta$ and $\vdash_n \Gamma, B \Rightarrow \Delta$;

(ix) $\vdash_n \Gamma \Rightarrow \forall x A, \Delta$ iff $\vdash_n \Gamma \Rightarrow A[x/y], \Delta$ $(y \notin \mathrm{FV}(\Gamma \cup \Delta \cup \{\forall x A\}))$;

(x) $\vdash_n \Gamma, \exists x A \Rightarrow \Delta$ iff $\vdash_n \Gamma, A[x/y], \Delta$ $(y \notin \mathrm{FV}(\Gamma \cup \Delta \cup \{\exists x A\}))$.

Proof. Straightforward by a tedious induction on the length of proofs. The verification is made easier by observing the complete symmetry between $+$, $\star$ and $\sqcap$, $\sqcup$ etc., and the fact that $A \multimap B$ behaves as ${\sim}A + B$. $\square$

♠ Exercise. Prove the inversion lemma.

3.8. Definition. A sequent $\Gamma \Rightarrow \Delta$ is said to be *primitive* if all formulas in Γ are of one of the forms ${\sim}A$, $A \multimap B$, $A \sqcap B$, $A + B$, $!A$, $?A$, $\forall x A$ or atomic but not $\mathbf{1}$, and all formulas in Δ are of one of the forms $A \sqcup B$, $A \star B$, $!A$, $?A$, $\exists x A$, or atomic but not $\mathbf{0}$. $\square$
Now the inversion lemma readily yields the following

Theorem. *Let* $\Gamma \Rightarrow \Delta$ *be derivable in* **CLL**. *Then there is a derivation* $\mathcal{D}$ *such that every subderivation* $\mathcal{D}'$ *of a non-primitive sequent* $\Gamma' \Rightarrow \Delta'$ *in* $\mathcal{D}$ *is obtained from subderivations* $\mathcal{D}_i$ *of primitive sequents* $\Gamma_i \Rightarrow \Delta_i$ *using exclusively applications of* $\mathrm{R}\multimap$, $\mathrm{R}+$, $\mathrm{R}\forall$, $\mathrm{R}\sqcap$, $\mathrm{R}{\sim}$, $\mathrm{R}\mathbf{0}$, $\mathrm{L}\star$, $\mathrm{L}\exists$, $\mathrm{L}\sqcup$, $\mathrm{L}\mathbf{1}$.
Proof. Suppose $\mathcal{D}^* \vdash \Gamma \Rightarrow \Delta$; we show by induction on $d(\mathcal{D}^*)$ that $\mathcal{D}^*$ can be transformed into a $\mathcal{D}^{**}$ with the required properties. Without loss of generality we may assume that $\mathcal{D}^*$ starts from atomic instances of Ax. Let $d(\mathcal{D}^*) = n$, and assume (induction hypothesis) that the required property has been established for all $\mathcal{D}$ starting

from atomic instances of the axioms with $d(\mathcal{D}) < n$. By the inversion lemma we can transform $\mathcal{D}^*$ into a $\mathcal{D}'$ with $d(\mathcal{D}') \leq n$ where $\Gamma \Rightarrow \Delta$ is obtained from primitive $\Gamma_i \Rightarrow \Delta_i$ (with subdeductions $\mathcal{D}_i$ by means of a (possibly empty) series of applications of R$\multimap$, R+, R$\forall$, R$\sqcap$, R$\sim$, R0, L$\star$, L$\exists$, L$\sqcup$, L1 only, while $d(\mathcal{D}_i) \leq n$. For any proper subderivation $\mathcal{D}'_i$ of $\mathcal{D}_i$ clearly $d(\mathcal{D}'_i) < n$ and the induction hypothesis may be applied; thus $\mathcal{D}'$ is transformed into $\mathcal{D}^{**}$. $\square$

Note that the theorem also applies to sublanguages of the full language $\{\star, +, \multimap, \sqcap, \sqcup, \sim, \mathbf{0}, \mathbf{1}, \top, \bot, !, ?, \forall, \exists\}$.

3.9. DEFINITION. (POS, NEG, CON). We define simultaneously *positive* (POS) and *negative* (NEG) contexts; CON = POS $\cup$ NEG. Let $P \in$ POS, $N \in$ NEG be arbitrary, and let A be an arbitrary formula; then POS and NEG are generated by the clauses

(i) $[\,], A \sqcap P, P \sqcap A, A \sqcup P, P \sqcup A, A \star P, P \star A, \forall x\, P, \exists x\, P,$
    $A \multimap P, !P, N \multimap A, \sim N \in$ POS;

(ii) $A \sqcap N, N \sqcap A, A \sqcup N, N \sqcup A, A \star N, N \star A, \forall x\, N, \exists x\, N,$
    $A \multimap N, !N, P \multimap A, \sim P \in$ NEG.

Clauses for $?, +$ may be added according to their definition in terms of $\sim, \star, !$. We write $F[A]$ if in $F[\,] \in$ CON the hole $[\,]$ is replaced by A. $\square$

N.B. Intuitively speaking, a context is nothing but a position of a subformula.

3.10. PROPOSITION. *(Monotonicity) Let $F[\,]$ be a context, and let $\vec{z}$ be the list of variables free in B,C but bound in $F[B], F[C]$. Then in $\mathbf{CLL_q}$ and $\mathbf{ILL_q}$*

(i) *If $F \in$ POS, then $\vdash\ \mathbf{1} \sqcap \forall \vec{z}(B \multimap C) \Rightarrow F[B] \multimap F[C]$,*

(ii) *If $F \in$ NEG, then $\vdash\ \mathbf{1} \sqcap \forall \vec{z}(B \multimap C) \Rightarrow F[C] \multimap F[B]$.*

In full $\mathbf{CLL}$, $\mathbf{ILL}$ we have only

(iii) *If $F \in$ POS, and $\vdash\ \Rightarrow (B \multimap C)$, then $\vdash\ \Rightarrow F[B] \multimap F[C]$,*

(iv) *If $F \in$ NEG, and $\vdash\ \Rightarrow (B \multimap C)$, then $\vdash\ \Rightarrow F[C] \multimap F[B]$.*

PROOF. Note that $\mathbf{1} \sqcap \forall x\, A$ is equivalent to $\forall x(\mathbf{1} \sqcap A)$, and that $\forall x \forall y\, A$ is equivalent to $\forall y \forall x\, A$. (i) and (ii) are proved by simultaneous induction on $F[\,]$, and similarly for (iii) and (iv). For (i) and (ii) we consider two typical cases.

Case 1. Let $F \equiv F'[\ ] \multimap A$, $F'[\ ] \in$ NEG. Then

$$\frac{\mathbf{1} \sqcap \forall \vec{x}(B \multimap C), F'[C] \Rightarrow F'[B]}{\dfrac{\mathbf{1} \sqcap \forall \vec{x}(B \multimap C), F'[C], F'[B] \multimap A \Rightarrow A}{\mathbf{1} \sqcap \forall \vec{x}(B \multimap C), F'[B] \multimap A \Rightarrow F'[C] \multimap A}}$$

Case 2. Let $F \equiv \forall x\, F'(x, [\ ]) \in$ POS. Then

$$\frac{\forall \vec{z}(\mathbf{1} \sqcap (B \multimap C)), F'(x, B) \Rightarrow F'(x, C)}{\dfrac{\forall \vec{z}(\mathbf{1} \sqcap (B \multimap C)), \forall x\, F'(x, B) \Rightarrow F'(x, C)}{\dfrac{\forall x \vec{z}(\mathbf{1} \sqcap (B \multimap C)), \forall x\, F'(x, B) \Rightarrow F'(x, C)}{\forall x \vec{z}(\mathbf{1} \sqcap (B \multimap C)), \forall x\, F'(x, B) \Rightarrow \forall x\, F'(x, C)}}}$$

For (iii)–(iv) consider

Case 3. $F[\ ] \equiv\ !F'[\ ] \in$ POS. By IH, if $\vdash \Rightarrow B \multimap C$, then $\vdash \Rightarrow F'[B] \multimap F'[C]$. Then $\vdash F'[B] \Rightarrow F'[C]$ implies $\vdash !F'[B] \Rightarrow F'[C]$ and this in turn implies $\vdash !F'[B] \Rightarrow\ !F'[C]$.

3.11. COROLLARY. *(Replacement of equivalents) Assume* $\vdash \Rightarrow B \multimap C$, $\vdash \Rightarrow C \multimap B$, *and let* $F[\ ]$ *be a context. Then*

(i) *If* $\vdash \Gamma, F[B] \Rightarrow \Delta$, *then* $\vdash \Gamma, F[C] \Rightarrow \Delta$.

(ii) *If* $\vdash \Gamma \Rightarrow F[B], \Delta$, *then* $\vdash \Rightarrow F[C], \Delta$.

The following proposition axiomatizes **CLL** relative to **ILZ**.

3.12. PROPOSITION. *If we add an axiom* $\sim\sim A \Rightarrow A$ *to* **ILZ**, *and define* $\sim A$ *as* $A \multimap \mathbf{0}$, *and* $?$, $+$, $\top$ *as de Morgan duals of* $!$, $\star$, $\bot$, *then all sequents of* **CLL** *with at most one formula in the succedent become derivable.*

PROOF. By induction on the length of deductions, we show for each derivable sequent $\Gamma \Rightarrow \Delta$ that we can prove $\Gamma \Rightarrow +\Delta$ in **ILZ** plus the extra axiom. Here $+(A_1, A_1, \ldots, A_n)$ abbreviates $A_1 + A_2 + \cdots + A_n$, and $\sim(A_1, \ldots, A_n)$ abbreviates $\sim A_1, \ldots, \sim A_n$. Consider e.g., the $\sim$-rule

$$\frac{\Gamma, A \Rightarrow \Delta}{\Gamma \Rightarrow \sim A, \Delta}$$

We must show in **ILZ** plus extra axiom that if

$$(1) \qquad\qquad\qquad \Gamma, A \Rightarrow +\Delta$$

then

$$\Gamma \Rightarrow \sim A, +\Delta$$

Now from (1) $\Gamma, A, \sim(+\Delta) \Rightarrow \Lambda$ (recall that Λ is the empty multiset), and hence $\Gamma, \sim\sim A, \sim(+\Delta) \Rightarrow \Lambda$ (cut with the axiom); therefore $\Gamma, \sim\sim A \star \sim(+\Delta) \Rightarrow \Lambda$, which by definition is $\Gamma, \sim(\sim A + (+\Delta)) \Rightarrow \Lambda$; hence $\Gamma \Rightarrow \sim\sim(\sim A + (+\Delta))$, so $\Gamma \Rightarrow \sim A + (+\Delta))$ (axiom), etc. $\square$

3.13. The one-sided calculus for CLL

The symmetry between the antecedent and the succedent part of sequents, introduced by the rules for $\sim$ in **CLL** permits a version of the sequent calculus with succedent formulas only. This is achieved by dropping $\sim$ as a logical operator. Instead, we regard the prime formulas as occurring in pairs $P, \sim P$, with $\sim\sim P$ literally identical with P by definition. For compound A, $\sim A$ is defined by de Morgan dualization:

$$\sim(A \sqcap B) := \sim A \sqcup \sim B, \quad \sim(A \star B) := \sim A + \sim B, \text{ etc.}$$

A sequent $\Gamma \Rightarrow \Delta$ of the two-sided calculus now corresponds to $\sim\Gamma, \Delta$ in the one-sided calculus (we drop $\Rightarrow$ from $\Rightarrow \sim\Gamma, \Delta$ since it has become redundant in the one-sided calculus). Each rule for a logical operation in the one sided calculus corresponds to two rules in the two-sided calculus; there are no rules for $\sim$. The rules are listed in Table 3.

3.14. Cut elimination

An application of the rule Cut is called a cut. A very important property of **CLL** and **ILL** is cut elimination; any sequent derivable with the help of the Cut rule is also derivable without Cut. In fact, Gentzen's original method, as described e.g., in Kleene 1952, also applies to **CLL** and **ILL**, with a slight twist. Proving cut elimination for the two-sided calculus for **CLL** yields a slightly stronger result than what is obtainable from cut elimination for the one-sided calculus. So we prefer to prove cut elimination for the two-sided calculus.

THEOREM. **CLL** and **ILL** *permit cut elimination.*
PROOF. We consider the two-sided calculus; this covers both **CLL** and **ILL**.

We summarize the method. The *rank* of a cut is the complexity of the cut formula. The *cutrank* of a deduction is the maximum rank of the cuts in the deduction. The *level* of a cut is the length of the subdeduction (i.e., the number of sequents in the subdeduction) ending in the cut. (Alternatively, one might define the level of a cut as

TABLE 3
The one-sided calculus for **CLL**

Axiom:
$$A, \sim A$$

Cut:
$$\frac{\Gamma, A \qquad \Delta, \sim A}{\Gamma, \Delta}$$

Rules for propositional operators:

$$\frac{\Gamma, A \quad \Delta, B}{\Gamma, \Delta, A \star B} \qquad \frac{\Gamma, A \quad \Gamma, B}{\Gamma, A \sqcap B} \qquad \frac{\Gamma, A}{\Gamma, A \sqcup B} \qquad \frac{\Gamma, B}{\Gamma, A \sqcup B}$$

$$\frac{\Gamma, A, B}{\Gamma, A + B} \qquad \frac{\Gamma}{\Gamma, 0} \qquad \Gamma, \top \qquad 1 \qquad \text{no rules for } \perp$$

Quantifier rules (y not free in Γ):

$$\frac{\Gamma, A[x/y]}{\Gamma, \forall x \, A} \qquad \frac{\Gamma, A[x/t]}{\Gamma, \exists x \, A}$$

Rules for exponentials:

$$\frac{\Gamma, ?A, ?A}{\Gamma, ?A} \qquad \frac{\Gamma}{\Gamma, ?A} \qquad \frac{\Gamma, A}{\Gamma, ?A} \qquad \frac{?\Gamma, A}{?\Gamma, !A}$$

---

the sum of the depths of the deductions of the premises.) As we shall see below, these notions have to be extended to some generalizations of the cut rule, but this is straightforward.

The proof of the eliminability of Cut proceeds by induction on the cutranks of proofs. For the induction step, we show how to replace a deduction with a single cut of rank n as its last rule by a deduction of the same conclusion and lower cutrank, by induction on the level of the terminal cut (i.e., by induction on the length of the deduction). Then we can lower the cutrank of an arbitrary deduction of cutrank n, by successively replacing subdeductions ending in a topmost cut of maximal degree by a new subdeduction of lower cutrank.

So let us consider a deduction with a single cut of maximal rank as its last rule. We regard the formula occurrence $!A$ obtained by contraction from two occurrences of $!A$ in an application of C!, and the formula occurrence $?B$ obtained by the contraction of two occurrences of $?B$ in an application of ?C as being *introduced* by these rule applications. Then for all applications of the rules or axioms for an operator or constant $\bowtie$ say, the formula occurrence of the form $\bowtie A$ (for a rule) or $\bowtie$ (for an axiom) introduced by the rule application is called the *principal* formula of the conclusion. In the treatment of a terminal cut we must distinguish two main cases. In the first main case the cut formula is principal in both premises of the terminal cut; in the second situation this is not the case.

First main case. In this case we replace the cut to be treated by zero or more cuts of the same rank, but of a lower level, and possibly some cuts of a lower rank. For example,

$$\frac{\dfrac{\Gamma \Rightarrow \Delta}{\Gamma, 1 \Rightarrow \Delta} \qquad \Rightarrow 1}{\Gamma \Rightarrow \Delta} \text{Cut}$$

becomes $\Gamma \Rightarrow \Delta$, and

$$\frac{\dfrac{\Gamma \Rightarrow A, \Delta \qquad \Gamma', B \Rightarrow \Delta'}{\Gamma, \Gamma', A \multimap B \Rightarrow \Delta, \Delta'} \text{L}\multimap \qquad \dfrac{\Gamma'', A \Rightarrow B, \Delta''}{\Gamma'' \Rightarrow A \multimap B, \Delta''} \text{R}\multimap}{\Gamma, \Gamma', \Gamma'' \Rightarrow \Delta, \Delta', \Delta''} \text{Cut}$$

becomes

$$\frac{\dfrac{\Gamma \Rightarrow A, \Delta \qquad \Gamma'', A \Rightarrow B, \Delta''}{\Gamma, \Gamma'' \Rightarrow B, \Delta''} \text{Cut} \qquad \Gamma', B \Rightarrow \Delta'}{\Gamma, \Gamma', \Gamma'' \Rightarrow \Delta, \Delta', \Delta''}$$

However, a complication arises in the following case. Consider a deduction of the following type, with $!A$ principal in both premises of the cut.

$$\frac{\dfrac{\mathcal{D}}{\dfrac{\Gamma, !A, !A \Rightarrow \Delta}{\Gamma, !A \Rightarrow \Delta}} \qquad \dfrac{\mathcal{D}'}{\dfrac{!\Gamma' \Rightarrow A, ?\Delta'}{!\Gamma' \Rightarrow !A, ?\Delta'}}}{\Gamma, !\Gamma' \Rightarrow \Delta, ?\Delta'} \text{Cut}$$

We are tempted to replace this by

$$\frac{\dfrac{\mathcal{D} \qquad \dfrac{\mathcal{D}'}{!\Gamma' \Rightarrow A, ?\Delta'}}{\dfrac{\Gamma, !A, !A \Rightarrow \Delta \qquad !\Gamma' \Rightarrow !A, ?\Delta'}{!A, \Gamma, !\Gamma' \Rightarrow \Delta, ?\Delta}} \text{Cut} \qquad \dfrac{\mathcal{D}'}{\dfrac{!\Gamma' \Rightarrow A, ?\Delta'}{!\Gamma' \Rightarrow !A, ?\Delta'}} \text{Cut}}{\dfrac{\Gamma, !\Gamma', !\Gamma' \Rightarrow \Delta, ?\Delta', ?\Delta'}{\Gamma, !\Gamma' \Rightarrow \Delta, ?\Delta'}} \text{C!,C?}$$

Here the double line indicates a number of successive contractions. Now above the topmost cut the sequent rank has decreased, but we have introduced a new cut below this. In order to avoid the difficulty for our induction, we also permit generalized Cut rules (which are derivable):

$$\frac{\Gamma \Rightarrow \,!A, \Delta \qquad \Gamma', (!A)^n \Rightarrow \Delta'}{\Gamma, \Gamma' \Rightarrow \Delta, \Delta'} \quad \text{Cut!}$$

$$\frac{\Gamma, ?A \Rightarrow \Delta \qquad \Gamma' \Rightarrow (?A)^n, \Delta'}{\Gamma, \Gamma' \Rightarrow \Delta, \Delta'} \quad \text{Cut?}$$

with $n > 1$, and where for any formula C, C^n denotes a multiset of n copies of C. We write Cut* for either Cut, Cut! or Cut? The notions of rank and level are extended in the obvious way. For C! and C? applications, if one of the occurrences of $!A$ on the left ($?A$ on the right) is principal, we regard the whole multiset of the occurrences on the left (right), deleted by the cut, as principal. The pairs of rules involving ! or ? and making the cut formula principal on both sides are (C!,R!), (L!,R!), (W!,R!), (L?,C?), (L?,R?), (L?,W?). We may restrict attention to the first three pairs, since the last three can be treated symmetrically.

Case of (C!,R!).

$$\frac{\dfrac{\overset{\mathcal{D}}{\Gamma, (!A)^{n+1} \Rightarrow \Delta}}{\Gamma, (!A)^n, \Rightarrow \Delta} \qquad \dfrac{\overset{\mathcal{D}'}{!\Gamma' \Rightarrow A, ?\Delta'}}{!\Gamma' \Rightarrow \,!A, ?\Delta'}}{\Gamma, !\Gamma' \Rightarrow \Delta, ?\Delta'} \quad \text{Cut*}$$

becomes

$$\frac{\overset{\mathcal{D}}{\Gamma, (!A)^{n+1} \Rightarrow \Delta} \qquad \dfrac{\overset{\mathcal{D}'}{!\Gamma' \Rightarrow A, ?\Delta'}}{!\Gamma' \Rightarrow \,!A, ?\Delta'}}{\Gamma, !\Gamma' \Rightarrow \Delta, ?\Delta'} \quad \text{Cut!}$$

Now the level of the cut has decreased.

Case of (L!,R!). Consider the deduction $\mathcal{D}^*$:

$$\frac{\dfrac{\overset{\mathcal{D}}{\Gamma, (!A)^n, A \Rightarrow \Delta}}{\Gamma, (!A)^{n+1} \Rightarrow \Delta} \qquad \dfrac{\overset{\mathcal{D}'}{!\Gamma' \Rightarrow A, ?\Delta'}}{!\Gamma' \Rightarrow \,!A, ?\Delta'}}{\Gamma, !\Gamma' \Rightarrow \Delta, ?\Delta'} \quad \text{Cut*}$$

In this case we must distinguish between two subcases: Cut* $\equiv$ Cut!, and Cut* $\equiv$ Cut. In the first subcase, the Cut* on !A is replaced by a Cut* on !A with a lower level, as follows.

$$
\cfrac{
\cfrac{
\mathcal{D} \qquad\qquad \mathcal{D}'
}{
\cfrac{\Gamma, (!A)^n, A \Rightarrow \Delta \qquad !\Gamma' \Rightarrow !A, ?\Delta'}{\Gamma, !\Gamma', A \Rightarrow \Delta, ?\Delta'} \text{Cut*} \qquad\qquad
\begin{array}{c}\mathcal{D}'\\[2pt] !\Gamma' \Rightarrow A, ?\Delta'\end{array}
}
}{
\cfrac{\Gamma, !\Gamma', !\Gamma' \Rightarrow ?\Delta', ?\Delta'}{\Gamma, !\Gamma' \Rightarrow ?\Delta'}
}
$$

where the double line indicates a succession of applications of C! and C?. In the second case, simply replace the $\mathcal{D}^*$ by

$$
\cfrac{
\begin{array}{cc}\mathcal{D} & \mathcal{D}'\end{array} \\
\Gamma, A \Rightarrow \Delta \qquad !\Gamma' \Rightarrow A, ?\Delta'
}{
\Gamma, \Gamma' \Rightarrow \Delta, ?\Delta'
}
$$

Case of (W!,R!). Here too we must distinguish between two subcases, analogous to, but slightly simpler than for the combination (L!,R!); this is left to the reader.

Second main case. The cut formula is not principal in at least one of the premises of the terminal cut. The idea is now to permute Cut* upwards over a premise where the cut formula is not principal. For example, the following piece of a deduction where the left premise of the Cut was obtained by an introduction rule for an additive constant

$$
\cfrac{
\cfrac{\Gamma, C \Rightarrow \Delta \quad \Gamma', C \Rightarrow \Delta'}{\Gamma'', C \Rightarrow \Delta''}\text{rule} \qquad \Gamma''' \Rightarrow C, \Delta'''
}{
\Gamma'', \Gamma''' \Rightarrow \Delta'', \Delta'''
}
$$

is transformed into

$$
\cfrac{
\cfrac{\Gamma, C \Rightarrow \Delta \quad \Gamma''' \Rightarrow C, \Delta'''}{\Gamma, \Gamma''' \Rightarrow \Delta, \Delta'''} \qquad \cfrac{\Gamma', C \Rightarrow \Delta' \quad \Gamma''' \Rightarrow C, \Delta'''}{\Gamma', \Gamma''' \Rightarrow \Delta', \Delta'''}
}{
\Gamma'', \Gamma''' \Rightarrow \Delta'', \Delta'''
}\text{rule}
$$

This works fine, except where the Cut* involved is in fact a Cut! or a Cut?, and the multiset $(!A)^n$ or $(?A)^n$ removed by the Cut! or Cut? is derived from *two* premises of a multiplicative rule (R⋆, L+, L⊸ or

Cut*); a representative example is

$$\frac{\dfrac{\Gamma,(!A)^p \Rightarrow \Delta, B \qquad \Gamma',(!A)^q \Rightarrow \Delta', C}{\Gamma,\Gamma',(!A)^{p+q} \Rightarrow, \Delta, \Delta', B \star C} \qquad \Gamma'' \Rightarrow !A, \Delta''}{\Gamma,\Gamma',\Gamma'' \Rightarrow \Delta, \Delta', \Delta'', B \star C}$$

where $p, q \geq 1$. (If either $p = 0$ or $q = 0$, there is no difficulty in permuting the Cut! upwards on the left.) In this case, cutting $\Gamma'' \Rightarrow !A, \Delta''$ with both the upper sequents on the left, followed by R$\star$ leaves us with duplicated Γ'', Δ''. To get out of this difficulty, we now look at the premise on the right. There are two possibilities. If $!A$ is not principal in the right hand premise of the cut, we can permute the Cut!-application upwards on the right. The obstacle which prevented permuting with the left hand premise does not occur here, since only a single occurrence of $!A$ is involved. On the other hand, if $!A$ is principal in the right hand premise, we must have $\Gamma'' \equiv !\Gamma''', \Delta'' \equiv ?\Delta'''$ for suitable Γ''', Δ''', and we may cut with the upper sequents on the left, followed by contractions of $!\Gamma''', !\Gamma'''$ into $!\Gamma'''$, and of $?\Delta''', ?\Delta'''$ into $?\Delta'''$ and an application of the multiplicative rule (R$\star$ in our example). $\square$

♠ EXERCISE. Complete the proof of cut elimination by considering the remaining cases.

REMARKS. (i) Basically the same idea works for the one-sided sequent calculus. There is a close relationship between the procedure in the two cases.

Let ϕ be the obvious map from two-sided sequents to one-sided sequents, i.e., $\phi(\Gamma \Rightarrow \Delta) = \sim\Gamma^*, \Delta^*$, where Γ^*, Δ^* are obtained by replacing implications $C \multimap D$ by $\sim C \sqcup D$. Translating proofs is straightforward, but note that the translation of an R$\sim$ or L$\sim$ step corresponds to a *repetition*. It turns out that the elementary steps for removing cuts, just mentioned, translate into the corresponding steps for the one-sided calculus — except where the cut formula is a negation, principal on both sides. It remains to be seen whether such a close correspondence remains when more complicated defined operators are treated as primitives in the two-sided calculus and as defined in the one-sided calculus.

(ii) The use of the generalized Cut rules Cut! and Cut? is similar to Gentzen's use of the rule Mix instead of Cut.

(iii) By Roorda (1989, improved 1991) it has been proved that even strong cut elimination holds, that is to say any strategy for removing cuts (subjected to some obvious restrictions as to the places where a cut elimination step may be applied) ultimately leads to a cut free proof.

3.15. As examples of easy applications of cut elimination we mention the following propositions.

PROPOSITION. *Deductions in* **CLL** *and* **ILL** *have the subformula property: if* $\vdash \Gamma \Rightarrow \Delta$ *is derivable, then there is a deduction containing subformulas of* Γ, Δ *only.*
PROOF. Immediate by considering cutfree proofs. $\square$

PROPOSITION. *A fragment of* **CLL** *determined by a subset* $\mathcal{L}$ *of* $\{\star, \multimap, \sqcap, \sqcup, \mathbf{1}, \bot, \top, \forall, \exists, !\}$ *is conservative over* **ILL** *(i.e., if* **CLL** *restricted to* $\mathcal{L}$ *proves* A, *then so does* **ILL** *restricted to* $\mathcal{L}$*) iff* $\mathcal{L}$ *does not include both* $\multimap$ *and* $\bot$.
PROOF. $\Leftarrow$ Suppose that $\bot$ is not in $\mathcal{L}$, let $\mathcal{D}$ be a cutfree proof of $\Gamma \Rightarrow A$, and assume that $\mathcal{D}$ contains a sequent with a consequent consisting of more than one formula. This can happen only if there is an application L$\multimap$ of the form

$$\frac{\Gamma \Rightarrow A, C \qquad \Gamma', B \Rightarrow}{\Gamma, \Gamma', A \multimap B \Rightarrow C}$$

We can then follow a branch in the deduction tree with empty succedents only. This branch must end in an axiom with empty succedent, which can only be **0**, which is excluded since the whole deduction is carried out in a sublanguage of **ILL**, or $\Gamma, \bot \Rightarrow$ which is excluded by assumption.

If $\multimap$ is not in $\mathcal{L}$, we can prove by a straightforward induction on the length of a cutfree deduction of a sequent $\Delta \Rightarrow B$, that all sequents in the deduction have a single consequent.

$\Rightarrow$ As to the converse, in the fragment $\{\multimap, \bot\}$ of **CLL** we can prove

$$P \multimap ((\bot \multimap Q) \multimap R), (P \multimap S) \multimap \bot \Rightarrow R \quad (P, Q, R, S \text{ atomic})$$

This sequent does not have a cutfree proof in **ILL**. $\square$

♠ EXERCISES.
1. Construct a **CLL**-derivation for the sequent mentioned in the proof.
2. Prove that for Γ in the language $\{\star, \multimap, \sqcap, \sqcup, \mathbf{1}, \bot, \top, \forall, \exists, !\}$ we have **CLL** $\vdash \Gamma \Rightarrow \bot$ iff **ILL** $\vdash \Gamma \Rightarrow \bot$ ('Glivenko's theorem for linear logic').

3.16. Decidability and undecidability

Recently some results on the decidability and undecidability of **CLL** and its subsystems have been obtained (Lincoln et. al. 1990a). In particular it has been shown that derivability of sequents in CLL_0 is pspace-complete, and that CLL_e is undecidable; see Chapter 20.

4   The calculus of two implications: a digression

The present chapter may be skipped by the reader, since it is not needed for understanding the later chapters. The material presented here, due to H. Schellinx, provides additional insight into the role of implication in **CLL**.

4.1.   Let us consider the following version of a sequent calculus for classical propositional logic, say $\mathbf{CL}_{\mathrm{if}}$ ("if" because the calculus is based on implication and falsehood).

Axioms $\qquad\qquad\qquad A \Rightarrow A \qquad\qquad \Gamma, \bot \Rightarrow \Delta$

Logical Rules:

$$\mathrm{R}{\to}\ \frac{\Gamma, A \Rightarrow B, \Delta}{\Gamma \Rightarrow A \to B, \Delta} \qquad \mathrm{L}{\to}\ \frac{\Gamma_0 \Rightarrow A, \Delta_0 \qquad \Gamma_1, B \Rightarrow \Delta_1}{\Gamma_0, \Gamma_1, A \to B \Rightarrow \Delta_0, \Delta_1}$$

Structural rules and Cut:

$$\mathrm{WL}\ \frac{\Gamma \Rightarrow \Delta}{\Gamma, B \Rightarrow \Delta} \qquad \mathrm{WR}\ \frac{\Gamma \Rightarrow \Delta}{\Gamma \Rightarrow B, \Delta}$$

$$\mathrm{CL}\ \frac{\Gamma, A, A \Rightarrow \Delta}{\Gamma, A \Rightarrow \Delta} \qquad \mathrm{CR}\ \frac{\Gamma \Rightarrow A, A, \Delta}{\Gamma \Rightarrow A, \Delta}$$

$$\text{Cut } \frac{\Gamma_0 \Rightarrow A, \Delta_0 \qquad \Gamma_1, A \Rightarrow \Delta_1}{\Gamma_0, \Gamma_1 \Rightarrow \Delta_0, \Delta_1}$$

We obtain full classical propositional logic by taking the other connectives as being defined in terms of $\rightarrow$, $\perp$. Weakening is essential in order to obtain the correct rules for defined $\vee$ and $\wedge$. It is also not difficult to see that $\mathbf{CL_{if}}$ permits elimination of Cut.

4.2. Additive implication

The rule for implication in $\mathbf{CL_{if}}$ is multiplicative, that is to say the rules are the same as for linear implication $\multimap$ in $\mathbf{CLL}$. Instead, we might have chosen the additive form of $\mathrm{L}{\rightarrow}$, i.e.,

$$\frac{\Gamma \Rightarrow A, \Delta \qquad \Gamma, B \Rightarrow \Delta}{\Gamma, A \rightarrow B \Rightarrow \Delta}$$

If we add to the calculus above the additive rules for implication and a R$\perp$ rule

$$\mathrm{R}\perp \frac{\Gamma \Rightarrow \Delta}{\Gamma \Rightarrow \perp, \Delta}$$

$$\mathrm{Ra1} \rightarrow \frac{\Gamma \Rightarrow B, \Delta}{\Gamma \Rightarrow A \rightarrow B, \Delta} \qquad \mathrm{Ra2} \rightarrow \frac{\Gamma, A \Rightarrow \Delta}{\Gamma \Rightarrow A \rightarrow B, \Delta}$$

$$\mathrm{La} \rightarrow \frac{\Gamma \Rightarrow A, \Delta \qquad \Gamma, B \Rightarrow \Delta}{\Gamma, A \rightarrow B \Rightarrow \Delta}$$

and we drop weakening and contraction, we obtain a calculus $\mathbf{CL_{if}^*}$ for which we can prove the following:

4.3. PROPOSITION. $\mathbf{CL_{if}^*}$ *is equivalent to* $\mathbf{CL_{if}}$ *but does not permit cut elimination.*

PROOF. From $\Gamma, A \Rightarrow \Delta$ we obtain $\Gamma, A \Rightarrow B, \Delta$:

$$\frac{\dfrac{\Gamma, A \Rightarrow \Delta}{\Gamma \Rightarrow A \rightarrow B, \Delta} \qquad \dfrac{A \Rightarrow A \qquad B \Rightarrow B}{A, A \rightarrow B \Rightarrow B}}{\Gamma, A \Rightarrow B, \Delta}$$

and similarly we obtain $\Gamma, A \Rightarrow B, \Delta$ from $\Gamma \Rightarrow B, \Delta$. So this establishes weakening. Moreover, if we take $\Gamma = \Lambda$, $\Delta = \{A\}$, $B = A$, we find $A \Rightarrow A, A$ and similarly $A, A \Rightarrow A$. From a Cut application

$$\frac{\Gamma, A, A \Rightarrow \Delta \qquad A \Rightarrow A, A}{\Gamma, A \Rightarrow \Delta}$$

we see that contraction holds. Finally we note that $P, Q \Rightarrow P$ for prime P, Q cannot be derived without Cut. $\square$

4.4. Obviously the derivability of structural rules is connected with the impossibility of eliminating cut; the derivation of weakening and contraction uses cut in combination with the fact that $\rightarrow$ obeys additive *and* multiplicative rules simultaneously, and for $\perp$ we have the "ex falso axiom" as well as R$\perp$ ("$\perp$-weakening") rule. So it is plausible that if we separate these two roles, we regain eliminability of cut.

Thus, if we

(a) split $\rightarrow$ into multiplicative $\multimap$, and additive $\rightsquigarrow$, and

(b) split $\perp$ into multiplicative **0** and additive $\perp$,

we obtain the following calculus:

4.5. The calculus CLL$_{\text{if}}$

Axioms
$$A \Rightarrow A \qquad \Gamma, \perp \Rightarrow \Delta \qquad \mathbf{0} \Rightarrow$$

Logical rules
$$\text{R0} \; \frac{\Gamma \Rightarrow \Delta}{\Gamma \Rightarrow \mathbf{0}, \Delta}$$

$$\text{R}\multimap \; \frac{\Gamma, A \Rightarrow B, \Delta}{\Gamma \Rightarrow A \multimap B, \Delta} \qquad \text{L}\multimap \; \frac{\Gamma_0 \Rightarrow \Delta_0, A \quad \Gamma_1, B \Rightarrow \Delta_1}{\Gamma_0, \Gamma_1, A \multimap B \Rightarrow \Delta_0, \Delta_1}$$

$$\text{R}\rightsquigarrow \; \frac{\Gamma \Rightarrow B, \Delta}{\Gamma \Rightarrow A \rightsquigarrow B, \Delta} \qquad \text{R}\rightsquigarrow \; \frac{\Gamma, A \Rightarrow \Delta}{\Gamma \Rightarrow A \rightsquigarrow B, \Delta}$$

$$\text{L}\rightsquigarrow \; \frac{\Gamma \Rightarrow A, \Delta \quad \Gamma, B \Rightarrow \Delta}{\Gamma, A \rightsquigarrow B \Rightarrow \Delta}$$

Cut rule
$$\text{Cut} \; \frac{\Gamma_0 \Rightarrow A, \Delta_0 \quad \Gamma_1, A \Rightarrow \Delta_1}{\Gamma_0, \Gamma_1 \Rightarrow \Delta_0, \Delta_1}$$

We can then prove

4.6. THEOREM. $\mathbf{CLL}_{if}$ *permits cut elimination and is equivalent to* $\mathbf{CLL}_0$.

PROOF. Cut elimination is straightforward and left as an exercise. To obtain the equivalence with $\mathbf{CLL}_0$, we use the following definitions:

$$
\begin{aligned}
A + B &:= (A \multimap \mathbf{0}) \multimap B \\
A \star B &:= (A \multimap (B \multimap \mathbf{0})) \multimap \mathbf{0} \\
A \sqcup B &:= (A \rightsquigarrow \perp) \rightsquigarrow B \\
A \sqcap B &:= (A \rightsquigarrow (B \rightsquigarrow \perp)) \rightsquigarrow \perp \\
\mathbf{1} &:= \mathbf{0} \multimap \mathbf{0} \\
\top &:= \perp \rightsquigarrow \perp \\
{\sim}A &:= A \multimap \mathbf{0}. \quad \square
\end{aligned}
$$

N.B. $A \rightsquigarrow B$ is provably equivalent to ${\sim}A \sqcup B$ or $(A \multimap \mathbf{0}) \sqcup B$.

♠ EXERCISE. Prove the theorem.

This theorem shows that $\mathbf{CLL}_0$ is a "logic of two arrows". However, there is only a single negation, as the next proposition shows.

4.7. PROPOSITION. *In* $\mathbf{CLL}_{if}$ *we can derive*

(i) $(A \multimap \mathbf{0}) \multimap \mathbf{0} \Leftrightarrow A$,

(ii) $(A \rightsquigarrow \perp) \rightsquigarrow \perp \Leftrightarrow A$,

(iii) $A \multimap \mathbf{0} \Leftrightarrow A \rightsquigarrow \perp$.

♠ EXERCISES.

1. Prove the proposition.

2. Show that $\Rightarrow A \rightsquigarrow A$ is not derivable in $\mathbf{CLL}_{if}$ and show that adding axioms of the form $\Rightarrow A \rightsquigarrow A$ to $\mathbf{CLL}_{if}$ is equivalent to adding an additive Cut rule to the calculus:

$$
\mathrm{Cut}_a \ \frac{\Gamma \Rightarrow A, \Delta \quad \Gamma, A \Rightarrow \Delta}{\Gamma \Rightarrow \Delta}
$$

5 Embeddings and approximations

5.1. Sources for this chapter are Girard 1987, Grishin 1974, Ono 1990a, Sambin 1989, and Schellinx 1990.

In removing the structural rules, but adding the exponentials instead, we have not lost anything: both classical and intuitionistic logic can be faithfully embedded into **CLL** as we shall see below.

Our intuition concerning the exponentials is that they permit us to apply weakening and contraction for the formulas to which they are applied; $!A$ means "use A any number of times in the antecedent" and $?A$ "use A any number of times in the succedent" (or "use $\sim A$ any number of times in the antecedent"). Using A either once or not at all in the antecedent (succedent) corresponds to using $\mathbf{1} \sqcap A$ ($\mathbf{0} \sqcup A$); and since in any given proof a formula is used only finitely often, we may expect that in a proof involving $!$ and $?$ these can be replaced by a proof in which only finite approximations $!_n$ and $?_n$ for suitable n are used, where

DEFINITION.

$$!_n A := \quad (\mathbf{1} \sqcap A) \star \cdots \star (\mathbf{1} \sqcap A) \quad (n \text{ times})$$

$$?_n A := (\mathbf{0} \sqcup A) + \cdots + (\mathbf{0} \sqcup A) \quad (n \text{ times}) \quad \square$$

Indeed we can prove

5.2. THEOREM. *(Approximation theorem for* **CLL***) Suppose that in the one-sided calculus for* **CLL** *we have shown* $\vdash \Gamma$, *and assume each occurrence* α *of* ! *in* Γ *to have been assigned a label* $n(\alpha) \in \mathbb{N} \setminus \{0\}$. *Then we can assign to each occurrence* β *of* ? *a label* $n(\beta) \in \mathbb{N} \setminus \{0\}$, *such that if* Γ' *is obtained from* Γ *by replacing everywhere any occurrence* α *of* ! *or* β *of* ? *by* $!_{n(\alpha)}, ?_{n(\beta)}$ *respectively, then* **CLL**$_q \vdash \Gamma'$.

PROOF. **CLL** $\vdash \Gamma \Rightarrow$ **CLL**$_q \vdash \Gamma'$ for a suitable labelling is proved by induction on the length of deductions in the one-sided calculus. That is to say, we prove by induction on n: if **CLL** $\vdash \Gamma$ by a deduction of length at most n, then for any labelling of the occurrences of ! in Γ there is a labelling of the ?-occurrences in Γ such that **CLL**$_q \vdash \Gamma'$ for the resulting Γ'.

Assume the induction hypothesis to hold for deductions of length at most n. Let $\vdash \Delta$ by a deduction of length $n + 1$. We distinguish cases according to the last rule applied; we treat three cases here and leave the others to the reader.

Case 1. The last rule applied is the $\sqcap$-rule: $\Delta \equiv \Gamma, A \sqcap B$ is obtained from Γ, A and Γ, B. Choose a labelling n of the !-occurrences in $\Gamma, A \sqcap B$; we suppose the !-occurrences in the premises Γ, A and Γ, B to be labelled correspondingly. By the induction hypothesis, we can choose a labelling m of the ?-occurrences in the first premise Γ, A and another labelling m' of the ?-occurrences in the second premise Γ, B such that the resulting approximations Γ', A' and Γ'', B'' are both derivable in **CLL**$_q$.

We note that all occurrences of subformulas $?C$ in the formulas of Γ are positive (since in the one-sided calculus $\sim$ applies only to prime formulas and $\multimap$ is not a primitive symbol). Now one readily sees that $?_p C \multimap ?_{p+q} C$, hence by monotonicity we may always replace an occurrence $?_p C$ by $?_{p+q} C$. So if the occurrences $?_{m(\beta)}$ in Γ' and $?_{m'(\beta)}$ in Γ'', both deriving from occurrence β of ? in Γ, are replaced by $?_{m(\beta)+m'(\beta)}$, for all ?-occurrences β in Γ, both are transformed into Γ''' such that Γ''', A' and Γ''', B'' are derivable, hence also **CLL**$_q \vdash \Gamma''', A' \sqcap B''$.

Case 2. Let $\Delta \equiv \Gamma, ?A$ be obtained from the premise $\Gamma, ?A, ?A$ by the contraction rule. Choose labels for the !-occurrences in $\Gamma, ?A$ and suppose $\Gamma, ?A, ?A$ to have been provided with the corresponding labels. By the induction hypothesis there is then a labelling of the ?-occurrences in $\Gamma, ?A, ?A$ such that the resulting $\Gamma', ?_n A', ?_m A''$ is derivable. (Since the labelling assigned to corresponding occurrences of ? in the two occurrences of A is not necessarily the same, we have designed the result by A' and A'' respectively. But as in the preceding case, we may replace corresponding occurrences $?_{p'}$ and

$?_{p''}$ in A' and A'' respectively by $?_{p'+p''}$ resulting in A'''; thus we find that $\Gamma', ?_{(n+m)}A'''$ is derivable.

Case 3. Let $\Delta \equiv ?\Gamma, !A$ be obtained from $?\Gamma, A$ by the !-rule. Suppose labels for the !-occurrences in Δ to have been chosen, where ! in $!A$ gets the label p. Let the !-occurrences in the premise $?\Gamma, A$ be correspondingly labelled (dropping the label p), and let us assume for notational simplicity that Γ consists of a single formula B. by the induction hypothesis there is a labelling of the ?-occurrences in $?B, A$ such that the resulting approximation $?_q B', A'$ is derivable by a deduction $\mathcal{D}'$. Weakening with formulas of the form $?_t C$ is always possible, hence if $\mathcal{D}$ is the deduction

$$\frac{\dfrac{1}{?_q B', 1} \qquad \dfrac{\mathcal{D}'}{?_q B', A'}}{?_q B', 1 \sqcap A'}$$

we obtain by the $\star$-rule

$$\frac{\dfrac{\mathcal{D}}{?_q B', 1 \sqcap A'} \qquad \dfrac{\mathcal{D}}{?_q B', 1 \sqcap A'}}{?_q B', ?_q B', !_2 A'}$$

which is equivalent to $?_{2q} B', !_2 A'$. Similarly we obtain, iterating the argument, that $?_{pq} B', !_p A'$ is derivable. $\square$

REMARK. In Girard, Scedrov & Scott 1990, a system of bounded linear logic is described, where subscripted $!_n$ appear as primitives of the calculus. The computational power turns out to be precisely that of the polynomial-time computable functions.

♠ EXERCISE. Formulate the approximation theorem for the two-sided sequent calculus.

For **ILL** we have a slightly different version.

5.3. DEFINITION. A *positive (negative)* occurrence of an operator ! in B is an occurrence arising by substituting in the hole of a positive (negative) context $F[\ ]$ a formula $!A$ such that $B \equiv F[!A]$. A *positive (negative)* occurrence of ! in $\Gamma \Rightarrow B$ is either a positive occurrence in the succedent B or a negative (positive) occurrence in one of the formula occurrences of Γ. $\square$

5.4. THEOREM. *(Approximation theorem for **ILL**)* Assume **ILL** $\vdash$ $\Gamma \Rightarrow A$, and suppose a label in $\mathbb{N} \setminus \{0\}$ has been assigned to each

TABLE 4

The Grishin embedding

(a) *The modified Grishin embedding.*

(a1) P^{+n} $:= \mathbf{0} \sqcup P,$ $\quad P^{-n}$ $:= \mathbf{1} \sqcap P$ for P prime,

(a2) $(\neg A)^{+n}$ $:= \sim(A^{-n}),$ $\quad (\neg A)^{-n}$ $:= \sim(A^{+n}),$

(a3) $(A \to B)^{+n} := A^{-n} \multimap B^{+n},$ $(A \to B)^{-n} := \sim A^{+n} \sqcup B^{-n},$

(a4) $(A \wedge B)^{+n}$ $:= A^{+n} \sqcap B^{+n},$ $(A \wedge B)^{-n}$ $:= A^{-n} \star B^{-n},$

(a5) $(A \vee B)^{+n}$ $:= A^{+n} + B^{+n},$ $(A \vee B)^{-n}$ $:= A^{-n} \sqcup B^{-n},$

(a6) $(\forall x\, A)^{+n}$ $:= \forall x(A^{+n}),$ $(\forall x\, A)^{-n}$ $:= (\forall x(A^{-n}))^{\mathbf{n}},$

(a7) $(\exists x\, A)^{+n}$ $:= \mathbf{n}(\exists x(A^{+n})),$ $(\exists x\, A)^{-n}$ $:= (\exists x(A^{-n})),$

where for all B

$$B^{\mathbf{1}} := B, \quad \mathbf{1}B := B, \quad B^{\mathbf{n}+1} := B^{\mathbf{n}} \star B, \quad (\mathbf{n}+1)B := \mathbf{n}B + B.$$

(b) *Grishin's embedding with exponentials.* Clause (b1)–(b5) are as (a1)–(a5), with $^{-}$ replacing $^{-n}$, $^{+}$ replacing $^{+n}$. (a6) and (a7) are replaced by

(b6) $(\forall x\, A)^{+}$ $:= \forall x(A^{+}),$ $(\forall x\, A)^{-}$ $:= \,!\forall x(A^{-}),$

(b7) $(\exists x\, A)^{+}$ $:= ?\exists x(A^{+})$ $(\exists x\, A)^{-}$ $:= \exists x(A^{-}),$

---

positive occurrence of ! in $\Gamma \Rightarrow A$. Then we can also assign a label to each negative occurrence of ! in $\Gamma \Rightarrow A$, such that if $\Gamma' \Rightarrow A'$ is obtained from $\Gamma \Rightarrow A$ by replacing occurrence α of ! with label $n(\alpha)$ by $!_{n(\alpha)}$, for all α, then $\mathbf{ILL_q} \vdash \Gamma' \Rightarrow A'$.

PROOF. Completely similar. $\square$

5.5. Grishin's embedding

We shall now discuss an embedding of **CL** into **CLL**. The idea for this embedding is found in Grishin 1974, and has been adapted to linear logic by Ono (1990a).

For the definition, see Table 4. There is an approximation variant (the a-clauses), defined in $\mathbf{CLL_q}$, and a version with exponentials (formulated by Ono, the b-clauses). We have the following

THEOREM.

$$\mathbf{CL} \vdash \Gamma \Rightarrow \Delta \ \text{ iff } \ \mathbf{CLL_q} \vdash \Gamma^{-n} \Rightarrow \Delta^{+n} \text{ for some } n,$$

$$\mathbf{CL} \vdash \Gamma \Rightarrow \Delta \ \text{ iff } \ \mathbf{CLL} \vdash \Gamma^{-} \Rightarrow \Delta^{+}.$$

The proof requires a number of lemmas.

LEMMA. *In* $\mathbf{CLL_q}$ *we have* $A^{-n} \Leftrightarrow 1 \sqcap A^{-n}, A^{+n} \Leftrightarrow 0 \sqcup A^{+n}$.

LEMMA. *In* $\mathbf{CLL_0}$ *we have*

$$A^- \Rightarrow A^{-(n+1)} \Rightarrow A^{-n} \Rightarrow A^{+n} \Rightarrow A^{+(n+1)} \Rightarrow A^+.$$

PROOF. By induction on the construction of A. $\square$

Proof of the embedding theorem for Grishin's translation.
The implications from right to left are straightforward, interpreting
$!A$ and $?A$ as A, $\star$ and $\sqcap$ as $\wedge$, $\sqcup$ and $+$ as $\vee$, $\multimap$ as $\rightarrow$, $\top$ and 1 as
$\top$ (or $P \rightarrow P$), $\perp$ and 0 as $\perp$ (or $A \wedge \neg A$), $\sim$ as $\neg$.

The direction from left to right is easily proved by induction on
the length of deductions, using the system for $\mathbf{CL}$ exhibited below.
(Weakening and Contraction are absorbed in the axioms. For this
system, an easy induction shows that $\Gamma, P \Rightarrow P, \Delta$ is derivable for
atomic P and arbitrary Γ, Δ, and a second induction yields $\Gamma, A \Rightarrow
A, \Delta$ for arbitrary Γ, Δ, A.)

Axioms $\qquad\qquad \Gamma, P \Rightarrow P, \Delta$ with all formulas atomic.

Rules

$$\frac{\Gamma \Rightarrow A, \Delta \qquad \Gamma, B \Rightarrow \Delta}{\Gamma, A \rightarrow B \Rightarrow \Delta} \qquad\qquad \frac{\Gamma, A \Rightarrow B, \Delta}{\Gamma \Rightarrow A \rightarrow B, \Delta}$$

$$\frac{\Gamma, A, B \Rightarrow \Delta}{\Gamma, A \wedge B \Rightarrow \Delta} \qquad\qquad \frac{\Gamma \Rightarrow A, \Delta \quad \Gamma \Rightarrow B, \Delta}{\Gamma \Rightarrow A \wedge B, \Delta}$$

$$\frac{\Gamma, A \Rightarrow \Delta \qquad \Gamma, B \Rightarrow \Delta}{\Gamma, A \vee B \Rightarrow \Delta} \qquad\qquad \frac{\Gamma \Rightarrow A, B, \Delta}{\Gamma \Rightarrow A \vee B, \Delta}$$

$$\frac{\Gamma \Rightarrow A, \Delta}{\Gamma, \neg A \Rightarrow \Delta} \qquad\qquad \frac{\Gamma, A \Rightarrow \Delta}{\Gamma \Rightarrow \neg A, \Delta}$$

$$\frac{\Gamma, \forall x\, A, A[x/t] \Rightarrow \Delta}{\Gamma, \forall x\, A \Rightarrow \Delta} \qquad\qquad \frac{\Gamma \Rightarrow A[x/y], \Delta}{\Gamma \Rightarrow \forall x\, A, \Delta}$$

$$\frac{\Gamma, A[x/y] \Rightarrow \Delta}{\Gamma, \exists x\, A \Rightarrow \Delta} \qquad\qquad \frac{\Gamma \Rightarrow A[x/t], \exists y\, A, \Delta}{\Gamma \Rightarrow \exists y\, A, \Delta}$$

(y not free in Γ, Δ)

We check the cases of the axioms and the quantifier rules. From

$$\frac{\Gamma, P \Rightarrow \Delta}{\Gamma, 1 \sqcap P \Rightarrow \Delta} \qquad \frac{\Gamma \Rightarrow \Delta}{\Gamma, 1 \Rightarrow \Delta} \qquad \frac{\Gamma \Rightarrow P, \Delta}{\Gamma \Rightarrow 0 \sqcup P, \Delta} \qquad \frac{\Gamma \Rightarrow \Delta}{\Gamma \Rightarrow \Delta, 0} \\ \qquad\qquad \frac{\Gamma \Rightarrow \Delta, 0}{\Gamma \Rightarrow \Delta, 0 \sqcup P}$$

we see that $\Gamma^{-n}, P^{-n} \Rightarrow P^{+n}, \Delta^{+n}$ is derivable in $\mathbf{CLL_q}$ starting from $P \Rightarrow P$.

Case L$\forall$. Assume

$$\Gamma^{-n}, (\forall x\, A^{-n})^{\mathbf{n}}, A^{-n}[x/t] \Rightarrow \Delta^{+n}$$

This yields

$$\Gamma^{-n}, (\forall x\, A^{-n})^{\mathbf{n}}, \forall x\, A^{-n} \Rightarrow \Delta^{+n}$$

hence with $A^{-(n+1)} \Rightarrow A^{-n}$ (lemma)

$$\Gamma^{-n}, (\forall x\, A^{-(n+1)})^{\mathbf{n}}, \forall x\, A^{-(n+1)} \Rightarrow \Delta^{+n}$$

Now we again use the lemma : $C^{-(n+1)} \Rightarrow C^{-n}$, $C^{+n} \Rightarrow C^{+(n+1)}$ and we find

$$\Gamma^{-(n+1)}, (\forall x\, A^{-(n+1)})^{\mathbf{n+1}} \Rightarrow \Delta^{+(n+1)}$$

Case R$\forall$. Assume

$$\Gamma^{-n} \Rightarrow A^{+n}[x/y], \Delta^{+n},$$

then $\Gamma^{-n} \Rightarrow \forall x\, A^{+n}, \Delta^{+n}$. $\square$

5.6. REMARKS

(i) Each instance of L$\forall$ or R$\exists$ increases the n by 1.
(ii) The translation of the Cut rule is

$$\frac{\Gamma_1^{-n}, A^{-n} \Rightarrow \Delta_1^{+n} \qquad \Gamma_2^{-n}, A^{+n} \Rightarrow \Delta_2^{+n}}{\Gamma_1^{+n}, \Gamma_2^{-n} \Rightarrow \Delta_1^{+n}, \Delta_2^{+n}}$$

but this cannot be justified by a schema

$$(*) \qquad \frac{\Gamma, A^{-n} \Rightarrow \Delta \qquad \Gamma' \Rightarrow A^{+n}, \Delta'}{\Gamma, \Gamma' \Rightarrow \Delta, \Delta'}$$

since this would require $A^{+n} \Rightarrow A^{-n}$ to hold. So $(*)$ is conservative for translated sequents, but not in general.
(iii) Other embeddings similar to Grishin's embedding with exponentials are discussed in Girard 1987.

TABLE 5
Definition of embedding of **IL** into **ILL** and **CLL**

P°	$:= P$ for P atomic,
$\perp^\circ$	$:= \perp,$
$(\neg A)^\circ$	$:= !A^\circ \multimap \perp,$
$(A \wedge B)^\circ$	$:= A^\circ \sqcap B^\circ,$
$(A \vee B)^\circ$	$:= !A^\circ \sqcup !B^\circ,$
$(A \to B)^\circ$	$:= !A^\circ \multimap B^\circ,$
$(\forall x\, A)^\circ$	$:= \forall x\, A^\circ,$
$(\exists x\, A)^\circ$	$:= \exists x !A^\circ$

5.7. Embedding of **IL** into **ILL** and **CLL**

The definition of the embedding is given in Table 5. The following theorem (Girard 1987) is easily proved:

THEOREM. **IL** $\vdash \Gamma \Rightarrow A$ iff **ILL** $\vdash \,!\Gamma^\circ \Rightarrow A^\circ$.

PROOF. For the direction from left to right we apply induction on the length of derivations. Translating a sequent $\Gamma \Rightarrow \Lambda$ as $!\Gamma^\circ \Rightarrow \perp$. The proof is then straightforward, using $!\forall x\, A \Rightarrow \forall x !A$, $!(A \sqcap B) \Rightarrow !A$, $!(A \sqcap B) \Rightarrow !B$.

For the other direction, replace in a deduction of $!\Gamma^\circ \Rightarrow A^\circ$ in **ILL** $!A$ by A, $\sqcup$ by $\vee$, $\star$ and $\sqcap$ by $\wedge$, $\multimap$ by $\to$, $\perp$ remains the same. Then the result is (modulo the interpolation of some steps and removing redundancies) a correct proof in **IL** of $\Gamma \Rightarrow A$. $\square$

N.B. The translations of instances of the Cut rule are valid!

5.8. Proving the corresponding result for the embedding $^\circ$ of **IL** into **CLL** requires more work. Note that from 3.15 we can rather easily see that for the $\perp$-free fragment the embedding is faithful: the translated proof of $!\Gamma^\circ \Rightarrow A^\circ$ is after cut elimination in fact a proof in **ILL**. However, the presence of $\perp$ causes trouble. By means of some extra considerations we can overcome this difficulty. The argument presented below is due to H. Schellinx.

DEFINITION. Let $\mathcal{F}$ be **CLL** restricted to $\sqcap$, $\sqcup$, $\multimap$, $\perp$, $\forall$, $\exists$, $!$ without Cut, and Ax restricted to atomic instances. Let us call a formula *simple* if it is either atomic, or of the form $A \sqcup B$ or of the form $\exists x A$. $\square$

We recall that (cf. 3.8) a sequent $\Gamma \Rightarrow \Delta$ in $\mathcal{F}$ is *primitive* if Γ contains only $-\!\circ$, $\sqcap$ $\perp$, $\forall$ and Δ only $\sqcup$, $!$, $\exists$. The theorem in 3.8 yields now that if $\vdash_n \Gamma \Rightarrow \Delta$ in $\mathcal{F}$, where $\vdash_n$ is defined as for the inversion lemma, then $\Gamma \Rightarrow \Delta$ is derivable from deductions of primitive sequents $\Gamma_i \Rightarrow \Delta_i$ of degree $\leq n$ using R$-\!\circ$, R$\forall$, R$\sqcap$, L$\exists$, L$\sqcup$ only.

If we have deduced $\Gamma \Rightarrow !\Delta, C, C$ not starting with $!$, the deduction looks like the picture below; in the dotted part only R$-\!\circ$, R$\forall$, R$\sqcap$, L$\exists$, L$\sqcup$ have been used, and the C_i are simple.

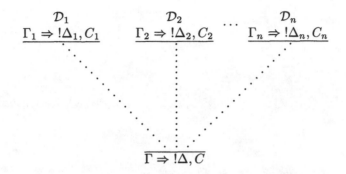

5.9. LEMMA. *Suppose we can derive in $\mathcal{F}$ either*

(a) $!\Gamma^\circ, \Pi^\circ \Rightarrow !\Delta^\circ$ *or*

(b) $!\Gamma^\circ, \Pi^\circ \Rightarrow !\Delta^\circ, B^\circ$ *with B° simple.*

Then we may assume that there is a deduction in $\mathcal{F}$ of (a) or (b) in which all sequents with more than one formula on the right side have one of the forms (1) or (2):

(1) $!\Sigma^\circ, \Delta^\circ \Rightarrow !\Theta^\circ, A^\circ$ *with $|\Theta| \geq 1$, A° simple,*

(2) $!\Sigma^\circ, \Delta^\circ \Rightarrow !\Theta^\circ$ *with $|\Theta| \geq 2$.*

PROOF. By induction on the measure of the derivation in $\mathcal{F}$ of (a) and (b). If in the proof of a sequent of type (a) ends with an R-rule, this must be R! with Π empty and $|\Delta| = 1$, so the proof ends with

$$\frac{!\Gamma^\circ \Rightarrow C^\circ}{!\Gamma^\circ \Rightarrow !C^\circ}$$

By the remark above, we may assume the derivation of $!\Gamma^\circ \Rightarrow C^\circ$ to be obtained applying R$-\!\circ$, R$\sqcap$, R$\forall$ to sequents $!\Gamma_i^\circ \Rightarrow C_i^\circ$, C_i° simple; use the induction hypothesis for the derivations of these sequents.

If a sequent of the form (b) has been derived by a proof terminating with an R-rule, the rule must have been either R$\sqcup$ or R$\exists$. In each of these cases we may apply the induction hypothesis for (a) to the derivation of the premise.

If a sequent of the form (a) or (b) has been derived with a final application of an L-rule in $\mathcal{F}$, the induction hypothesis may be applied to the derivations of the premises. For example, if the last rule is $\multimap$L, the deduction ends with

$$\frac{!\Gamma_1^\circ, \Pi_1^\circ \Rightarrow !A^\circ, !\Delta_1^\circ, C^\circ \qquad !\Gamma_2^\circ, \Pi_2^\circ, B^\circ \Rightarrow !\Delta_2^\circ}{!\Gamma_1^\circ, !\Gamma_2^\circ, \Pi_1^\circ, \Pi_2^\circ, !A^\circ \multimap B^\circ \Rightarrow !\Delta_1^\circ, !\Delta_2^\circ, C^\circ}$$

or with C° in the succedent of the second premise instead. If a sequent of type (a) or (b) is an axiom there is nothing to prove. $\square$

5.10. LEMMA. *If $\mathcal{F} \vdash !\Gamma^\circ \Rightarrow A^\circ$, then we may assume that all applications of R$\multimap$, R$\forall$ in the derivation have a single formula in the succedent.*

PROOF. We may assume $!\Gamma^\circ \Rightarrow A^\circ$ to have been obtained from $!\Gamma_i^\circ \Rightarrow A_i^\circ$, A_i° primitive, by use of R$\sqcap$, R$\forall$, R$\multimap$ only. By the preceding lemma, we may assume all sequents with more than one formula occurrence in the succedent in derivations of $!\Gamma_i^\circ \Rightarrow A_i^\circ$ to have the form (i) or (ii) mentioned in the lemma. Obviously such sequents can never arise as conclusion of R$\multimap$ or R$\forall$. $\square$

5.11. THEOREM. *The embedding of Table 5 is faithful, i.e.,*

$$\mathbf{IL} \vdash \Gamma \Rightarrow A \text{ iff } \mathbf{CLL} \vdash !\Gamma^\circ \Rightarrow A^\circ.$$

PROOF. Consider any proof in $\mathcal{F}$ of $!\Gamma^\circ \Rightarrow A^\circ$, with only singleton succedents in any application of R$\multimap$ or R$\forall$. Erase !, replace $\sqcup$, $\sqcap$, $\multimap$ by $\vee$, $\wedge$, $\rightarrow$ respectively and we have obtained a proof of $\Gamma \Rightarrow A$ in the sequent calculus exhibited below.

Axioms: $\qquad\qquad\qquad A \Rightarrow A \qquad \Gamma, \bot \Rightarrow A$

Rules:

$$\text{R} \wedge \frac{\Gamma \Rightarrow A, \Delta \qquad \Gamma \Rightarrow B, \Delta}{\Gamma \Rightarrow A \wedge B, \Delta} \qquad \text{L} \wedge_i \frac{\Gamma, A_i \Rightarrow \Delta}{\Gamma, A_0 \wedge A_1 \Rightarrow \Delta}$$

$$\text{R} \vee_i \frac{\Gamma \Rightarrow A_i, \Delta}{\Gamma \Rightarrow A_0 \vee A_1, \Delta} \qquad \text{L} \vee \frac{\Gamma, A \Rightarrow \Delta \qquad \Gamma, B \Rightarrow \Delta}{\Gamma, A \vee B \Rightarrow \Delta}$$

$$R \to \frac{\Gamma, A \Rightarrow B}{\Gamma \Rightarrow A \multimap B} \qquad L \to \frac{\Gamma \Rightarrow A, \Delta \quad \Gamma', B \Rightarrow \Delta'}{\Gamma, \Gamma', A \to B \Rightarrow \Delta, \Delta'}$$

$$R\forall \frac{\Gamma \Rightarrow A}{\Gamma \Rightarrow \forall y\, A[x/y]} \qquad L\forall \frac{\Gamma, A[x/t] \Rightarrow \Delta}{\Gamma, \forall x\, A \Rightarrow \Delta}$$

$$R\exists \frac{\Gamma \Rightarrow A[x/t], \Delta}{\Gamma \Rightarrow \exists x\, A, \Delta} \qquad L\exists \frac{\Gamma, A \Rightarrow \Delta}{\Gamma, \exists y\, A[x/y] \Rightarrow \Delta}$$

It is easy to see that this calculus is equivalent to a more usual Gentzen calculus, where everywhere the succedents contain a single formula (if $\Gamma \Rightarrow B_1, \ldots, B_n$ is derivable in this calculus, then $\Gamma \Rightarrow B_1 \vee \cdots \vee B_n$ is derivable in the usual calculus for intuitionistic predicate logic). $\square$

REMARK. As noted by Schellinx, the result may also be derived as a corollary of the well-known faithfulness of Gödel's embedding of intuitionistic propositional logic in the modal logic S4 (a result going back to McKinsey & Tarski 1948; cf. Flagg & Friedman 1986).

5.12. Embedding of CLL into ILZ. This can be done by a translation which is nothing but a variant of the well known Kolmogorov translation of classical logic into intuitionistic logic. The definition is exhibited in Table 6.

TABLE 6

The Kolmogorov-style translation

DEFINITION. We define A^k by induction on the complexity of A. We let $\sim A := A \multimap 0$.

(1) P^k $:= \sim\sim P$ for P prime, not 0;

(2) 0^k $:= 0$;

(3) $(A \circ B)^k$ $:= \sim\sim(A^k \circ B^k)$ for $\circ \in \{\star, \sqcap, \multimap\}$;

(4) $(A + B)^k$ $:= \sim(\sim A^k \star \sim B^k)$;

(5) $(A \sqcup B)^k$ $:= \sim(\sim A^k \sqcap \sim B^k)$;

(6) $(!A)^k$ $:= \sim\sim !A^k$;

(7) $(?A)^k$ $:= \sim ! \sim A^k$;

(8) $(\forall x\, A)^k$ $:= \sim\sim \forall x A^k$;

(9) $(\exists x\, A)^k$ $:= \sim \forall x \sim A^k$.

THEOREM. $\mathbf{CLL} \vdash \Gamma \Rightarrow A$ *iff* $\mathbf{ILZ} \vdash \Gamma^k \Rightarrow A^k$.

PROOF. By induction on the complexity of A one establishes $\vdash \sim\sim A^k \circ\!\!-\!\!\circ A^k$. Then one proves by induction on the length of deductions that all deductions in $\mathbf{CLL}$, axiomatized as $\mathbf{ILZ}$ with axioms $\sim\sim A \multimap A$ added, translate into deductions into $\mathbf{ILZ}$. This is done by showing that any rule application under the translation corresponds to a (derived) rule. We consider two typical cases.

Case 1. Assume $\Gamma, A \Rightarrow C$ and $\Gamma, B \Rightarrow C$. Then

$$
\cfrac{
\cfrac{
\cfrac{
\cfrac{\Gamma, A \Rightarrow C \quad \mathbf{0} \Rightarrow}{\Gamma, \sim C, A \Rightarrow}}{\Gamma, \sim C, A \Rightarrow \mathbf{0}}}{\Gamma, \sim C \Rightarrow \sim A}
\qquad
\cfrac{
\cfrac{
\cfrac{\Gamma, B \Rightarrow C \quad \mathbf{0} \Rightarrow}{\Gamma, \sim C, B \Rightarrow}}{\Gamma, \sim C, B \Rightarrow \mathbf{0}}}{\Gamma, \sim C \Rightarrow \sim B}
}{
\cfrac{
\cfrac{\Gamma, \sim C \Rightarrow \sim A \sqcap \sim B \qquad \mathbf{0} \Rightarrow}{\Gamma, \sim C, \sim(\sim A \sqcap \sim B) \Rightarrow}}{\Gamma, \sim(\sim A \sqcap \sim B) \Rightarrow \sim\sim C}
}
$$

Thus if we have $\Gamma^k, A^k \Rightarrow C^k$ and $\Gamma^k, B^k \Rightarrow C^k$ it follows that $\Gamma^k, (A \sqcup B)^k \Rightarrow C^k$ since $\sim\sim C^k \circ\!\!-\!\!\circ C^k$.

Case 2. Suppose $\Gamma \Rightarrow A$, then

$$
\cfrac{
\cfrac{
\cfrac{
\cfrac{\Gamma \Rightarrow A \qquad \mathbf{0} \Rightarrow}{\Gamma, \sim A \Rightarrow}}{\Gamma, \sim A \sqcap \sim B \Rightarrow}}{\Gamma, \sim A \sqcap \sim B \Rightarrow \mathbf{0}}}{\Gamma \Rightarrow \sim(\sim A \sqcap \sim B)}
$$

which shows that from $\Gamma^k \Rightarrow A^k$ we obtain $\Gamma^k \Rightarrow (A \sqcup B)^k$. $\square$

6 Natural deduction systems for linear logic

6.1. The natural deduction systems presented here serve here, firstly, as useful intermediate systems, and secondly, as illustrations of the formula-as-types idea from the introduction. Table 7 lists the rules for the natural deduction systems N-**CLL** and N-**ILL**. In the names of the rules, "I" stands for "Introduction", and "E" for "Elimination". !E^w means "weakening elimination of !", and !E^c "contraction elimination of !". Note that the !E-rules, by specializing $\Gamma \equiv \,!B$, $\Delta \equiv \Gamma$ yield

$$\frac{\Gamma \vdash A}{\Gamma, !B \vdash A} \qquad \frac{\Gamma, B \vdash A}{\Gamma, !B \vdash A} \qquad \frac{\Gamma, !B, !B \vdash A}{\Gamma, !B \vdash A}$$

Conversely, from these special cases the general rules are easily derived.

6.2. LEMMA. *In the natural deduction systems:*

(i) *If* $\Gamma \vdash A$ *and* $A, \Gamma' \vdash B$ *then* $\Gamma, \Gamma' \vdash B$.

(ii) *If* $\Gamma, A \vdash B$ *and* $\Gamma' \vdash \sim B$ *then* $\Gamma, \Gamma' \vdash \sim A$.

(iii) *If* $\Gamma \vdash A$ *then* $\Gamma \vdash \sim\sim A$.

(iv) *If* $\Gamma, A \vdash B$ *then* $\Gamma, \sim B \vdash \sim A$.

♠ EXERCISE. Prove the lemma.

TABLE 7
Natural deduction systems for linear logic

Ax $A \vdash A$ TI $\Gamma \vdash \top$ $\bot$E $\Gamma, \bot \vdash A$

$\star$I $\dfrac{\Gamma \vdash A \quad \Delta \vdash B}{\Gamma, \Delta \vdash A \star B}$ $\star$E $\dfrac{\Gamma \vdash A \star B \quad \Delta, A, B \vdash C}{\Gamma, \Delta \vdash C}$

$\sqcap$I $\dfrac{\Gamma \vdash A \quad \Gamma \vdash B}{\Gamma \vdash A \sqcap B}$ $\sqcap$E$_i$ $\dfrac{\Gamma \vdash A_0 \sqcap A_1}{\Gamma \vdash A_i}$ $(i \in \{0, 1\})$

$\sqcup$I$_i$ $\dfrac{\Gamma \vdash A_i}{\Gamma \vdash A_0 \sqcup A_1}$ $(i \in \{0, 1\})$ $\sqcup$E $\dfrac{\Delta \vdash A \sqcup B \quad A, \Gamma \vdash C \quad B, \Gamma \vdash C}{\Gamma, \Delta \vdash C}$

$\multimap$I $\dfrac{\Gamma, A \vdash B}{\Gamma \vdash A \multimap B}$ $\multimap$E $\dfrac{\Gamma \vdash A \multimap B \quad \Delta \vdash A}{\Gamma, \Delta \vdash B}$

1I $\vdash \mathbf{1}$ 1E $\dfrac{\Gamma \vdash \mathbf{1} \quad \Delta \vdash A}{\Gamma, \Delta \vdash A}$

0-rule $\dfrac{\Gamma, A \multimap \mathbf{0} \vdash \mathbf{0}}{\Gamma \vdash A}$ (N-**CLL** only)

$\exists$I $\dfrac{\Gamma \vdash A[x/t]}{\Gamma \vdash \exists x\, A}$ $\exists$E $\dfrac{\Gamma \vdash \exists x\, A \quad \Delta, A[x/y] \vdash C}{\Gamma, \Delta \vdash C}$

$\forall$I $\dfrac{\Gamma \vdash A[x/y]}{\Gamma \vdash \forall x\, A}$ $\forall$E $\dfrac{\Gamma \vdash \forall x\, A}{\Gamma \vdash A[x/t]}$

!I $\dfrac{!\Gamma \vdash A}{!\Gamma \vdash !A}$!E^w $\dfrac{\Gamma \vdash !B \quad \Delta \vdash A}{\Gamma, \Delta \vdash A}$

!E $\dfrac{\Gamma \vdash !B \quad \Delta, B \vdash A}{\Gamma, \Delta \vdash A}$!E^c $\dfrac{\Gamma \vdash !B \quad \Delta, !B, !B \vdash A}{\Gamma, \Delta \vdash A}$

---

t free for x, y free for x and not free in A.

For N-**CLL**, $+$ and ? are defined by de Morgan duality; $\sim A$ is defined as $A \multimap \mathbf{0}$.

6.3. DEFINITION. A (natural deduction-) *interpretation* of a sequent $A_1, \ldots, A_n \Rightarrow B_1, \ldots, B_m$ is any sequent

$$A_1, \ldots, A_n, \sim B_1, \ldots, \sim B_{i-1}, \sim B_{i+1}, \ldots, \sim B_m \vdash B_i \ (1 \leq i \leq m) \text{ or}$$

$$A_1, \ldots, A_{i-1}, A_{i+1}, \ldots, A_n, \sim B_1, \ldots, \sim B_m \vdash \sim A_i \ (1 \leq i \leq n). \quad \square$$

6.4. THEOREM. $\Gamma \Rightarrow \Delta$ *is provable in* **CLL** *iff an interpretation of* $\Gamma \Rightarrow \Delta$ *is provable in* N-**CLL**.

For both **CLL** and **ILL** : $\Gamma \Rightarrow A$ *is provable in the sequent calculus iff* $\Gamma \Rightarrow A$ *is provable in natural deduction.*

PROOF. From N-**CLL** to **CLL** : use induction. The axioms and introduction rules immediately translate into axioms and R-rules of **CLL**. As to E-rules: suppose e.g.,

$$\Gamma' \Rightarrow A \star B, \quad \Gamma'', A, B \Rightarrow C;$$

from the second $\Gamma'', A \star B \Rightarrow C$; apply Cut. If $\Gamma \Rightarrow A \sqcap B$, apply Cut with $A \sqcap B \Rightarrow A$ etc.

From **CLL** to N-**CLL** : we consider a cut-free proof in **CLL**. We have to show that there is an interpretation in N-**CLL** provable. For axioms this is immediate. Suppose $\Gamma', A \vdash B$ (interpretation of $\Gamma, A \Rightarrow \Delta$). Then $\Gamma', \sim B \vdash \sim A$ by

$$\frac{\dfrac{\Gamma', A \vdash B \qquad B \multimap 0 \vdash B \multimap 0}{\Gamma', A, \sim B \vdash 0}}{\Gamma', \sim B \vdash \sim A}$$

etc. The proof is entirely routine. $\square$

6.5. Terms for natural deduction proofs of intuitionistic logic

We shall now associate terms with the sequents in natural deduction proofs of intuitionistic propositional logic, as an introduction to a similar term system for N-**ILL**$_e$. We stick to a sequent notation; open assumptions correspond to free variables (a variable may be thought of as representing a hypothetical unspecified proof of the assumption). We treat $\Gamma, \Delta, \ldots$ as *sets* (not multisets) of type statements of the form

$$x_1 : A_1, \ldots, x_n : A_n$$

with $A_1, \ldots, A_n$ a multiset of formulas, the x_i all distinct.

$$\text{Ax} \quad x : A \Rightarrow x : A \qquad \bot\text{-Ax} \quad \Gamma, x : \bot \Rightarrow E^{\bot}(x) : A$$

$$\text{C} \; \frac{\Gamma, x : A, y : A \Rightarrow t_i : B}{\Gamma, z : A \Rightarrow t[x, y/z, z] : B} \qquad \text{W} \; \frac{\Gamma \Rightarrow t : B}{\Gamma, z : A \Rightarrow t : B}$$

$$\wedge\text{I} \; \frac{\Gamma \Rightarrow s : A \quad \Gamma \Rightarrow t : B}{\Gamma \Rightarrow \langle s, t \rangle : A \wedge B} \qquad \wedge\text{E}_i \; \frac{\Gamma \Rightarrow t : A_0 \wedge A_1}{\Gamma \Rightarrow \pi_i t : A_i}$$

$$\rightarrow\text{I} \; \frac{\Gamma, x : A \Rightarrow t : B}{\Gamma \Rightarrow \lambda x.t : A \rightarrow B} \qquad \rightarrow\text{E} \; \frac{\Gamma \Rightarrow t : A \rightarrow B \quad \Gamma \Rightarrow s : A}{\Gamma \Rightarrow E^{\rightarrow}(t, s) : B}$$

$$\vee\text{I}_i \; \frac{\Gamma \Rightarrow t : A_i}{\Gamma \Rightarrow \kappa_i t : A_0 \vee A_1} \qquad \vee\text{E} \; \frac{\Gamma \Rightarrow t : A \vee B \quad \Gamma, y : B \Rightarrow s' : C \quad \Gamma, x : A \Rightarrow s : C}{\Gamma \Rightarrow E^{\vee}_{x,y}(t, s, s') : C}$$

$E^{\bot}$, $E^{\rightarrow}$, $E^{\vee}$ are the elimination operators for $\bot, \rightarrow, \vee$ respectively. $E^{\rightarrow}$ is usually called application and $E^{\rightarrow}(t, s)$ is abbreviated as $t(s)$ or even ts. The subscripts x, y in $E^{\vee}_{x,y}(t, s, s')$ indicate that x is bound in s and y is bound in s'. An alternative notation would be $E^{\vee}(t, (x)s, (y)s')$, where $(x), (y)$ are used to indicate binding without associating it with applications of the $\rightarrow$I-rule. For the $\wedge$E-rules we can give an alternative form, making the pattern of elimination constants more uniform:

$$\frac{\Gamma \Rightarrow t : A_0 \wedge A_1 \quad \Delta, x : A_i \Rightarrow s : C}{\Gamma, \Delta \Rightarrow E^{\wedge,i}_x(t, s) : C}$$

One may take $E^{\wedge,i}_x(t, s) = s[x/\pi_i t]$.

The term calculus exhibited above is a Curry-style type assignment calculus, in which the terms themselves are untyped; that is to say our statements "$t : A$" as exhibited above may be read as "t (untyped) can be assigned type A". In this chapter here we are more interested in a variant with rigid types, in which the terms with all their subtypes are unambiguously typed, e.g.,

$$\frac{\Gamma, x : A \Rightarrow t : B}{\Gamma \Rightarrow \lambda x : A.t : A \rightarrow B} \qquad \Gamma \Rightarrow E^{\bot}_A(x) : A$$

In particular the elimination constants are in need of extra typing information. We write e.g., $E^{\rightarrow}(t : A \rightarrow B, s : A)$ or $(t : A \rightarrow B)(s : A)$, etc. It is routine to supplement the notation in this way; for the rigidly typed version the terms completely code the proof trees (this is not the case for the Curry-version, why?). However, in order not to encumber our notation too much, we shall use the calculus exhibited above also as "shorthand" for the calculus with rigid typing. For computational aspects (i.e., contractions on proof trees corresponding to conversions on terms, see below), there is little difference between the two versions and one can just as well study the Curry-style system.

Usually (cf. Troelstra & van Dalen 1988, section 10.8) the rules C and W are omitted, that is to say the (free and bound) variables in the terms refer to actually used assumptions in the proof trees, and identifications between assumptions of the same formula in the proof tree by labeling them with the same variable are made in advance, not as a step in the construction of the proofterm. Compare, for example, the following two versions of a proof of $F \equiv (A \to (A \to B)) \to (A \to B)$:

$$\frac{\dfrac{y : A \to (A \to B) \quad x : A}{yx : A \to B} \quad x : A}{\dfrac{(yx)x : B}{\dfrac{\lambda x.(yx)x : A \to B}{\lambda yx.(yx)x : F}}}$$

$$\frac{\dfrac{\dfrac{y : A \to (A \to B) \quad x : A}{yx : A \to B} \quad x' : A}{\dfrac{(yx)x' : B \quad (x, x' \mapsto z)}{\dfrac{\lambda x.(yx)x : A \to B}{\lambda yx.(yx)x : F}}}}{}$$

In the right hand tree the renaming and identification of x and x' is a separate step.

"Cut elimination" for natural deductions is the removal of detours, i.e., contraction of introductions immediately followed by eliminations. Thus

$$\frac{\dfrac{\Gamma, x : A \Rightarrow t : B}{\Gamma \Rightarrow \lambda x.t : A \to B} \quad \Gamma \Rightarrow s : A}{\Gamma \Rightarrow (\lambda x.t)(s) : B} \qquad \text{contracts to} \quad \Gamma \Rightarrow t[x/s] : B$$

i.e., $(\lambda x.t)(s)$ is contracted to $t[x/s]$ (β-conversion), and

$$\frac{\dfrac{\Gamma \Rightarrow s_0 : A_0 \quad \Gamma \Rightarrow s_1 : A_1}{\Gamma \Rightarrow \langle s_0, s_1 \rangle : A_0 \wedge A_1}}{\Gamma \Rightarrow \pi_i \langle s_0, s_1 \rangle : A_i} \qquad \text{contracts to} \quad \Gamma \Rightarrow s_i : A_i$$

i.e., $\pi_i \langle s, t \rangle$ contracts to s_i etc. (We shall not enter in the complications of the so-called "permutation conversions" for $\vee$; cf. Troelstra & van Dalen 1988, section 10.8, or Girard, Lafont & Taylor 1988.)

In the case of the sequent calculus it is also possible to assign terms to the formulas in the deduction. However, in this case the terms do not uniquely code the sequent proof, but a natural deduction proof associated in an obvious way with the sequent calculus proof. For example,

$$
\dfrac{\dfrac{\mathcal{E}}{C,\Gamma,B \Rightarrow E} \quad \dfrac{\mathcal{E}'}{\Gamma' \Rightarrow A}}{\dfrac{C,\Gamma,\Gamma',A \to B \Rightarrow E}{C \wedge D,\Gamma,\Gamma',A \to B \Rightarrow E}}
\qquad \text{and} \qquad
\dfrac{\dfrac{\dfrac{\mathcal{E}}{C,\Gamma,B \Rightarrow E}}{C \wedge D,\Gamma,B \Rightarrow E} \quad \dfrac{\mathcal{E}'}{\Gamma' \Rightarrow A}}{C \wedge D,\Gamma,\Gamma',A \to B \Rightarrow E}
$$

both represent the same natural deduction proof of the form

$$
\dfrac{A \to B \quad \dfrac{\begin{array}{c}\mathcal{D}'\\ A\end{array}}{[B]} \quad \dfrac{C \wedge D}{[C]}}{\begin{array}{c}\mathcal{D}\\ E\end{array}}
$$

where $\mathcal{D}'$ corresponds to the proof $\mathcal{E}'$ of $\Gamma' \Rightarrow A$, and $\mathcal{D}$ to the proof $\mathcal{E}$ of $C,\Gamma,B \Rightarrow E$. The sequent calculus encodes a certain order of application of the rules which is irrelevant from the viewpoint of the natural deduction calculus. the sequent rules with term-assignment coincide for the R-rules with the I-rules, except that contexts are added, e.g.,

$$
\dfrac{\Gamma \Rightarrow s : A \quad \Gamma \Rightarrow t : B}{\Gamma \Rightarrow \langle s,t \rangle : A \wedge B}
$$

The Cut rule becomes substitution:

$$
\dfrac{\Gamma \Rightarrow s : A \quad x : A,\Delta \Rightarrow t : A}{\Gamma,\Delta \Rightarrow t[x/s] : A} \ \text{Cut}
$$

whereas the L-rules involve special substitutions:

$$
\wedge\mathrm{L}_i \ \dfrac{\Gamma,x : A_i \Rightarrow t : B}{\Gamma,z : A_0 \wedge A_1 \Rightarrow t[x/\pi_i z] : B}
$$

$$
\to\mathrm{L} \ \dfrac{\Gamma \Rightarrow s : A \quad x : B,\Delta \Rightarrow t : C}{\Gamma,\Delta,z : A \to B \Rightarrow t[x/zs] : C}
$$

$$
\vee\mathrm{L} \ \dfrac{\Gamma,x : A \Rightarrow s : C \quad \Gamma,y : B \Rightarrow t : C}{\Gamma,z : A \vee B \Rightarrow \mathrm{E}^{\vee}_{x,y}(z,s,t) : C}
$$

6.6. Term notation for N-ILL$_e$

We are now ready to present a term calculus for N-**ILL**$_e$; the result is given in Table 8. Again, this calculus may be seen either as a Curry-style type assignment calculus, or as shorthand for a rigidly typed system in which all terms with their subterms are unambiguously typed; in the latter case the terms are just notations for deductions, and from the term associated with the bottom sequent in a derivation the entire derivation may be read off.

As in the case of intuitionistic logic, transformations of proof trees contracting an introduction followed by an elimination (of the same logical operator) correspond to contractions on terms.

♠ EXERCISE. Formulate the contraction rules on terms for N-**ILL**$_e$.

6.7. Linear lambda-terms (digression)

If we restrict attention to the implicational fragment of N-**ILL** and use typed lambda-terms to denote the proof trees, it is readily seen that all typed λ-terms denoting deductions have the property that in every subterm $\lambda x.t$, x occurs exactly once in t. Such terms we call linear (typed) λ-terms.

We may also consider the untyped linear λ-terms, i.e., untyped λ-terms satisfying the same restriction. As shown by Hindley (1989), such terms are always Curry-typable. To be precise, we need some definitions.

6.8. DEFINITION. *Type schemes* (α, β, γ) are built from type variables (a, b, c) by means of $\rightarrow$, i.e., if α, β are type schemes, then so is $\alpha \rightarrow \beta$. A *type-assignment statement* is an expression $t : \alpha$, α a type scheme, t an untyped λ-term.

Type schemes are assigned to λ-terms by the rules $\rightarrow$e and $\rightarrow$i:

$$
\begin{array}{cc}
\dfrac{\begin{array}{cc}\mathcal{D} & \mathcal{D}' \\ t:\alpha\rightarrow\beta & t':\alpha\end{array}}{tt':\beta} \;\rightarrow\text{e}
&
\dfrac{\begin{array}{c}[x:\alpha] \\ \mathcal{D} \\ t:\beta\end{array}}{\lambda x.t:\alpha\rightarrow\beta} \;\rightarrow\text{i}
\end{array}
$$

In the rule $\rightarrow$i x must not occur free in an assumption open above $t : \beta$, except in an assumption $x : \alpha$. If $\mathcal{B} \equiv x_1 : \alpha_1, \ldots, x_n : \alpha_n$ for distinct x_i, then

$$\mathcal{B} \vdash t : \beta$$

means that $t : \beta$ is derivable from $\mathcal{B}$ by means of $\rightarrow$i and $\rightarrow$e. □

TABLE 8
Term assignment for N-**ILL**$_e$

$$x : A \Rightarrow x : A \qquad \Gamma \Rightarrow 1 : \top \qquad \Gamma, x : \bot \Rightarrow E^{\bot}(x) : A$$

$$\frac{\Gamma \Rightarrow s : A \quad \Delta \Rightarrow t : B}{\Gamma, \Delta \Rightarrow s \star t : A \star B} \qquad \frac{\Gamma \Rightarrow s : A \star B \quad \Delta, x : A, y : B \Rightarrow t : C}{\Gamma, \Delta \Rightarrow E^{\star}_{x,y}(s,t) : C}$$

$$\frac{\Gamma \Rightarrow s : A \quad \Gamma \Rightarrow t : B}{\Gamma \Rightarrow \langle s,t \rangle : A \sqcap B} \qquad \frac{\Gamma \Rightarrow s : A_0 \sqcap A_1}{\Gamma \Rightarrow \pi_i s : A_i}$$

$$\frac{\Gamma \Rightarrow s : A_i}{\Gamma \Rightarrow \kappa_i s : A_0 \sqcup A_1} \qquad \frac{\Delta \Rightarrow s : A \sqcup B \quad \Gamma, y : B \Rightarrow t'' : C}{\Gamma, \Delta \Rightarrow E^{\sqcup}_{x,y}(s,t',t'') : C}$$

with $\Gamma, x : A \Rightarrow t' : C$ above.

$$\frac{\Gamma, x : A \Rightarrow t : B}{\Gamma \Rightarrow \lambda x.t : A \multimap B} \qquad \frac{\Gamma \Rightarrow s : A \multimap B \quad \Delta \Rightarrow t : A}{\Gamma, \Delta \Rightarrow st : B}$$

$$\Rightarrow * : 1 \qquad \frac{\Gamma \Rightarrow s : 1 \quad \Delta \Rightarrow t : A}{\Gamma, \Delta \Rightarrow E^{\mathbf{1}}(s,t) : A}$$

$$\frac{!\Gamma \Rightarrow t : A}{!\Gamma \Rightarrow !t : !A} \qquad \frac{\Gamma \Rightarrow s : !B \quad \Delta, x : B \Rightarrow t : A}{\Gamma, \Delta \Rightarrow E^{!}_{x}(s,t) : A}$$

$$\frac{\Gamma \Rightarrow s : !B \quad \Delta \Rightarrow t : A}{\Gamma, \Delta \Rightarrow E^{w}(s,t) : A} \qquad \frac{\Gamma \Rightarrow s : !B \quad \Delta, x : !B, y : !B \Rightarrow t : A}{\Gamma, \Delta \Rightarrow E^{c}_{x,y}(s,t) : A}$$

Γ, Δ sets of statements $x_i : A_i$ with the x_i all distinct; Γ, Δ disjoint.

6.9. DEFINITION. *t has a type scheme* α if, for some $\mathcal{B}$, $\mathcal{B} \vdash t : \alpha$, and *t is stratified* iff it has a type scheme. $\square$
Then we have

6.10. THEOREM. *Every linear term is stratified.*
We shall not give the proof here. Arbitrary λ-terms are certainly not always stratified.

A similar result can be stated for terms constructed from the combinators B, C, I which have types

$$B : (\alpha \to \beta) \to ((\beta \to \gamma) \to (\alpha \to \gamma)),$$
$$C : (\alpha \to (\beta \to \gamma)) \to (\beta \to (\alpha \to \gamma)),$$
$$I : \alpha \to \alpha.$$

This corresponds to the implicational fragment of the Hilbert-system for **ILL** (see next chapter). The corresponding logic is called BCI-logic. The same holds for the slightly stronger BCK-logic, where axiom I has been replaced by

$$K : \alpha \to (\beta \to \alpha).$$

BCK- and BCI-logic have many more interesting properties (e.g., the 2-property or 2-1-property: every provable formula of BCI-logic is obtained by substitution in a provable formula in which every propositional variable occurs exactly twice; see end of Chapter 17).

7 Hilbert-type systems

7.1. Our next aim is to give a Hilbert-type axiomatization of linear logic (i.e., an axiomatization comparable to the well-known axiomatizations of classical logic based on modus ponens and generalization as the only rules). We use the natural deduction systems of the preceding chapter as an intermediate step in proving the equivalence between the sequent calculi and the formalisms H-**ILL** and H-**CLL**. The main source for this chapter is Avron 1988. For the systems concerned see Table 9.

7.2. LEMMA. *All axioms of Table 9 except (4) are provable in* **ILL** *(i.e.,* **ILL** $\vdash \Rightarrow F$ *for each instance F of an axiom); (4) is provable in* **CLL**.
PROOF. Straightforward. $\Box$

7.3. LEMMA. H-**ILL** $\vdash (A \star B \multimap C) \multimap (A \multimap (B \multimap C))$.
PROOF. Put $D \equiv (A \star B \multimap C) \multimap (B \multimap C)$. Then

(1) $$B \multimap (A \star B \multimap D)$$

(2) $$A \multimap (B \multimap A \star B) \text{ (axiom (5))}$$

(3) $$[A \multimap (B \multimap (A \star B))] \multimap [((B \multimap A \star B) \multimap D) \multimap (A \multimap D)]$$

again by axiom B,

65

(4) $((B \multimap A \star B) \multimap D) \multimap (A \multimap D)$ ($\multimap$-Rule, (2), (3))

(5) $A \multimap ((A \star B \multimap C) \multimap (B \multimap C))$ ($\multimap$-Rule, (1), (4))

(6) $(A \star B \multimap C) \multimap (A \multimap (B \multimap C))$

by an application of axiom C, (5) and $\multimap$-Rule. $\square$

7.4. DEFINITION. (Deducibility from assumptions in the H-systems)
As before, in $\Gamma \vdash A$, Γ is a finite multiset.

   A deduction from assumptions is constructed in tree form as follows. At the top appear assumptions $A \vdash A$ or $\vdash B$ with B an axiom
of the list. Deduction trees $\mathcal{D}$, $\mathcal{D}'$ are combined into new deduction
trees by means of the rules, i.e., the following are again deduction
trees (x not free in C, Γ):

$$
\begin{array}{cc}
\mathcal{D} \quad\quad \mathcal{D}' \\
\dfrac{\Gamma \vdash A \quad \Gamma' \vdash A \multimap B}{\Gamma, \Gamma' \vdash B}
\end{array}
\qquad
\begin{array}{cc}
\mathcal{D} \quad\quad \mathcal{D}' \\
\dfrac{\Gamma \vdash A \quad \Gamma \vdash B}{\Gamma \vdash A \sqcap B}
\end{array}
$$

$$
\begin{array}{c}
\mathcal{D} \\
\dfrac{\Gamma \vdash C \multimap A}{\Gamma \vdash C \multimap \forall x \, A}
\end{array}
\qquad
\begin{array}{c}
\mathcal{D} \\
\dfrac{\Gamma \vdash A \multimap C}{\Gamma \vdash \exists x \, A \multimap C}
\end{array}
\qquad
\begin{array}{c}
\mathcal{D} \\
\dfrac{\vdash A}{\vdash \,!A}
\end{array}
$$

where $\overset{\mathcal{D}}{\Gamma \vdash} B$ indicates a deduction tree with conclusion $\Gamma \vdash B$.

   So we shall say that $\Gamma \vdash A$ is derivable iff there is a deduction tree
ending with conclusion $\Gamma \vdash A$. $\square$

7.5. REMARKS

(i) In the $\multimap$-fragment, in a deduction of $\Gamma \vdash A$, each formula of Γ is
used exactly once.
(ii) Note that the ordinary rule of generalization: " If $\Gamma \vdash A$, then
$\Gamma \vdash \forall x \, A \ (x \notin \mathrm{FV}(\Gamma))$" is obtainable as a special case. By $1 \multimap (A \multimap$
$A)$ also $A \multimap (1 \multimap A)$ (use axiom C), hence $\Gamma \vdash 1 \multimap A$ by modus
ponens; so $\Gamma \vdash 1 \multimap \forall x \, A$, and by $\vdash 1$ and modus ponens $\Gamma \vdash \forall x \, A$.

7.6. LEMMA. *The deduction theorem holds for the H-systems, i.e.,*

$$\Gamma, A \vdash B \ \Rightarrow \Gamma \vdash A \multimap B$$

where $\vdash$ is deducibility in the H-systems.
PROOF. By induction on the length of derivations. To get an appropriate formulation of the induction hypothesis (IH), we take

TABLE 9
Hilbert type systems for linear logic

Axioms for H-**CLL** *and* H-**ILL** :

(1) $A \multimap A$ (axiom I)

(2) $(A \multimap B) \multimap ((B \multimap C) \multimap (A \multimap C))$ (axiom B)

(3) $(A \multimap (B \multimap C)) \multimap (B \multimap (A \multimap C))$ (axiom C)

(4) $((A \multimap \mathbf{0}) \multimap \mathbf{0})) \multimap A$ (for H-**CLL**$_q$ and H-**CLL** only)

(5) $A \multimap (B \multimap A \star B)$

(6) $(A \multimap (B \multimap C)) \multimap (A \star B \multimap C)$

(7) $\mathbf{1}$

(8) $\mathbf{1} \multimap (A \multimap A)$

(9) $A \sqcap B \multimap A, \; A \sqcap B \multimap B$

(10) $(A \multimap B) \sqcap (A \multimap C) \multimap (A \multimap B \sqcap C)$

(11) $A \multimap A \sqcup B, \; B \multimap A \sqcup B$

(12) $(A \multimap C) \sqcap (B \multimap C) \multimap (A \sqcup B \multimap C)$

(13) $A \multimap \top$

(14) $\bot \multimap A$

(15) $\forall x \, A \multimap A[x/t]$

(16) $A[x/t] \multimap \exists x \, A$

Rules:

$\multimap$-Rule $A, \; A \multimap B \Rightarrow B$ (Modus ponens)

$\sqcap$-Rule $A, \; B \Rightarrow A \sqcap B$ (Adjunction)

$\forall$-Rule $B \multimap A \Rightarrow B \multimap \forall x \, A \; (x \notin \mathrm{FV}(B))$

$\exists$-Rule $A \multimap B \Rightarrow \exists x \, A \multimap B \; (x \notin \mathrm{FV}(B))$

Additional axioms for H-**CLL** *and* H-**ILL** :

(17) $B \multimap (!A \multimap B)$

(18) $(!A \multimap (!A \multimap B)) \multimap (!A \multimap !B)$

(19) $!(A \multimap B) \multimap (!A \multimap !B)$

(20) $!A \multimap A$

(21) $!A \multimap !!A$

Additional rule:

!-Rule $A \Rightarrow !A$

For all deductions of length k with conclusion $\Gamma \vdash B$, if $A \in \Gamma$, then $\Gamma \setminus \{A\} \vdash A \multimap B$.

where $\Gamma \setminus \{A\}$ means the multiset Γ with an occurrence of A left out. Note that for a deduction ending with $\vdash B$ there is nothing to prove.

For $k = 1$ the truth of the IH is immediate. For the induction step we distinguish cases according to the last rule applied.

Case 1. Suppose $\Gamma, A \vdash B$ is obtained by modus ponens from $\Gamma' \vdash C \multimap B, \Gamma'' \vdash C$.

Case 1a. $\Gamma' \equiv \Delta, A$; by induction hypothesis $\Delta \vdash A \multimap (C \multimap B)$; apply modus ponens with axiom C to obtain $\Delta \vdash C \multimap (A \multimap B)$; with $\Gamma'' \vdash C$ we get $\Gamma \vdash A \multimap B$.

Case 1b. If $\Gamma'' \equiv \Delta, A$, then by IH $\Delta \vdash A \multimap C$; combination with axiom B and $\Gamma' \vdash C \multimap B$ yields $\Gamma \vdash A \multimap B$.

Case 2. Suppose $\Gamma, A \vdash B \sqcap C$ has been obtained by the adjunction rule from $\Gamma, A \vdash B$ and $\Gamma, A \vdash C$. Then the induction hypothesis yields $\Gamma \vdash A \multimap B$, $\Gamma \vdash A \multimap C$, and with the adjunction rule $\Gamma \vdash (A \multimap B) \sqcap (A \multimap C)$. With axiom (10) follows $\Gamma \vdash A \multimap (B \sqcap C)$.

Case 3. Suppose $\Gamma, A \vdash B \multimap \forall x\, C$ was obtained from $\Gamma, A \vdash B \multimap$ C by the $\forall$-rule. By IH $\Gamma \vdash A \multimap (B \multimap C)$, hence $\Gamma \vdash A \star B \multimap C$ by axiom (6), and hence $\Gamma \vdash A \star B \multimap \forall x\, C$ by the $\forall$-rule. By Lemma 7.3 then also $\Gamma \vdash A \multimap (B \multimap \forall x C)$.

Case 4. Suppose $\Gamma, A \vdash \exists x\, C \multimap B$ was obtained from $\Gamma, A \vdash C \multimap$ B by the E-rule; IH yields $\Gamma \vdash A \multimap (C \multimap B)$, hence with axiom C, $\Gamma \vdash C \multimap (A \multimap B)$; E-rule yields $\Gamma \vdash \exists x\, C \multimap (A \multimap B)$; again by axiom C, $\Gamma \vdash A \multimap (\exists x\, C \multimap B)$.

Case 5. $\vdash\ !A$ is obtained from $\vdash A$ by the !-Rule: there is nothing to prove. $\square$

7.7. Remark. The rule $!\Gamma \vdash A \Rightarrow !\Gamma \vdash !A$ is a derived rule for the H-systems. For if $!\Gamma \equiv !C_1, \ldots, !C_n$, and $!\Gamma \vdash A$ we find by the deduction theorem

$$\vdash\ !C_1 \multimap !C_2 \multimap \cdots !C_n \multimap A$$

hence with the !Rule

$$\vdash\ !(!C_1 \multimap !C_2 \multimap \cdots !C_n \multimap A)$$

so with (19)

$$\vdash\ !!C_1 \multimap !(!C_2 \multimap \cdots !C_n \multimap A),$$

hence by (21) $\vdash\ !C_1 \multimap !(!C_2 \multimap \cdots !C_n \multimap !A)$, etc. until we have

$$\vdash\ !C_1 \multimap !C_2 \multimap \cdots !C_n \multimap !A$$

and then by modus ponens $!\Gamma \vdash !A$.

7.8. THEOREM. *For* **S** $\equiv$ **CLL** *or* **ILL**:

$$\mathbf{S} \vdash \Gamma \Rightarrow A \text{ iff } \Gamma \vdash_{\text{H-}\mathbf{S}} A.$$

PROOF. We use the fact that

$$\mathbf{S} \vdash \Gamma \Rightarrow A \text{ iff N-}\mathbf{S} + \Gamma \vdash A.$$

We show by induction on the length of a deduction tree in N-**S** that if $\Gamma \vdash A$ in N-**S**, then $\Gamma \vdash_{\text{H-}\mathbf{S}} A$.

Each rule application in N-**S** can be replaced by axioms and the five rules of the H-systems. We discuss some typical cases.

Case 1. Assume $\Gamma'', A, B \vdash C$ in H-**S**, then with the deduction theorem $\Gamma'' \vdash A \multimap (B \multimap C)$, with axiom (6) $\Gamma'' \vdash A \star B \multimap C$, and if $\Gamma' \vdash A \star B$ in H-**S**, then $\Gamma', \Gamma'' \vdash C$.

Case 2. Suppose $\Gamma, A \vdash C$, $\Gamma, B \vdash C$ in H-**S**, then $\Gamma \vdash A \multimap C$, $\Gamma \vdash B \multimap C$ and with adjunction $\Gamma \vdash (A \multimap C) \sqcap (B \multimap C)$, hence with axiom (12) $\Gamma \vdash A \sqcup B \multimap C$.

Case 3. Suppose $\Gamma, !B, !B \vdash A$; with the deduction theorem $\Gamma \vdash !B \multimap (!B \multimap A)$, and with axiom (18) $\Gamma \vdash !B \multimap A$; so with modus ponens and $!B \vdash !B$, it follows that $\Gamma, !B \vdash A$. Etc. $\square$

7.9. Deduction from hypotheses; theories

In sequent calculi for linear logic we can distinguish several notions of deduction from hypotheses.

DEFINITION. B is an *internal consequence* of the multiset $\Gamma \equiv A_1, \ldots, A_n$, iff the sequent $\Gamma \Rightarrow B$ is derivable, and B is an *external consequence* of Γ iff $\Rightarrow B$ is derivable from hypotheses $\Rightarrow A_1$, $\ldots, \Rightarrow A_n$ (each used once). Notation: $\Gamma \vdash^e B$. $\square$

Clearly, the two notions differ: $C \sqcap \mathbf{1}$ is an external consequence of C, but not an internal consequence of C. Using Cut we see that external consequence is weaker than internal consequence.

REMARKS. (i) If $\Gamma \vdash^e B$ is derivable without use of L$\sqcup$, R$\sqcap$, L?, R!, then also $\vdash \Gamma \Rightarrow B$; and if $!\Gamma \vdash^e B$ is derivable without use of L$\sqcup$, R$\sqcap$, then also $\vdash !\Gamma \Rightarrow B$.

(ii) We may extend the notion of external consequence to sequents; a sequent S is *externally deducible* from a multiset $\mathcal{X}$ of sequents (notation $\mathcal{X} \vdash^e S$) if S can be deduced with the elements of $\mathcal{X}$ (each used once) appearing as axioms.

For any sequent $S \equiv A_1, \ldots, A_n \Rightarrow B$, write S^* for $A_1 \multimap (A_2 \multimap \ldots (A_n \multimap B) \ldots)$. If $\mathcal{X} \equiv S_1, \ldots S_n$, then $\mathcal{X} \vdash^e S$ iff $S_1^*, \ldots, S_n^* \vdash^e S^*$, as is readily verified.

Of more importance than external deducibility is the notion of deducibility in a theory.

DEFINITION. A theory **T** is a set of sequents, and a sequent S is *derivable in* **T** if S can be derived from sequents in **T** used arbitrarily often as axioms. Notation: **T** $\vdash$ S. □

PROPOSITION. *Let* **T** $\equiv \{A_1, \ldots, A_n\}$ *be a finite theory, and* $S \equiv \Gamma \Rightarrow B$ *a sequent; then* $T \vdash S$ *iff* $\vdash\, !A_1, \ldots, !A_n, \Gamma \Rightarrow B$.

8 Algebraic semantics

8.1. Algebraic semantics for linear logic, as presented here, is for linear logic what Boolean-valued models are for classical logic, and Heyting-valued models for intuitionistic logic. So far there are no interesting applications except one result in Chapter 12, but nevertheless algebraic semantics does provide some insight and may serve as an introduction to the categorical semantics of later chapters. The exposition below is based on Ono 1990b. For **CLL**, a proof was already given by Girard (1987). See also Abrusci 1990, Ono 1989, Sambin 1989.

8.2. DEFINITION. $\mathcal{X} \equiv (X, \sqcap, \sqcup, \bot, \multimap, \star, \mathbf{1})$ is an *IL-algebra* (*intuitionistic linear algebra*) if

   (i) $(X, \sqcap, \sqcup, \bot)$ is a lattice with bottom $\bot$;

   (ii) $(X, \star, \mathbf{1})$ is a commutative monoid with unit $\mathbf{1}$;

   (iii) if $x \leq x'$, $y \leq y'$, then $x \star y \leq x' \star y'$ and $x' \multimap y \leq x \multimap y'$;

   (iv) $x \star y \leq z$ iff $x \leq y \multimap z$.

An *IL-morphism* from $\mathcal{X}$ to $\mathcal{Y}$ is a map from X to Y preserving the operations and the constant $\mathbf{1}$. An IL-algebra with an additional constant $\mathbf{0}$ (zero) is called an *IL-algebra with zero* or *ILZ-algebra*. We write

$$\sim x := x \multimap \mathbf{0}$$

71

An IL-algebra with zero is a *CL-algebra* (*classical linear algebra*) if

(v) $x = \sim\sim x$ for all x.

A *complete* IL-algebra (ILZ-algebra, CL-algebra) is an IL-algebra (ILZ-algebra, CL-algebra) which is complete as a lattice. $\square$

REMARK. Viewing posets as special case of categories, (iii) expresses that $\star : \mathcal{X} \times \mathcal{X} \longrightarrow \mathcal{X}$ and $\multimap: \mathcal{X}^{\mathrm{op}} \times \mathcal{X} \longrightarrow \mathcal{X}$ are functors. (iv) expresses that the functor $- \star y$ is left adjoint to $y \multimap -$; so $- \star y$ preserves all colimits.

8.3.   LEMMA. *In any IL-algebra* $\mathcal{X} \equiv (X, \sqcap, \sqcup, \bot, \multimap, \star, \mathbf{1})$, *for all* x, y, z *in* X:

(i) $z \star (x \sqcup y) = (z \star x) \sqcup (z \star y)$ *and moreover, if the join* $\bigsqcup_{i \in I} y_i$ *exists, then* $x \star \bigsqcup_{i \in I} y_i = \bigsqcup_{i \in I} (x \star y_i)$.

(ii) $x \multimap (y \multimap z) = x \star y \multimap z$.

(iii) $\bot \multimap \bot$ *is top of* X,

and in any CL-algebra

(iv) $x \sqcup y = \sim(\sim x \sqcap \sim y)$, $x \multimap y = \sim(x \star \sim y)$.

PROOF.
(i) Using clauses (i) and (iv) in the definition of IL-algebra, $z \star (x \sqcup y) \leq v$ iff $x \sqcup y \leq z \multimap v$ iff $(x \leq z \multimap v)$ and $(y \leq z \multimap v)$ iff $(x \star z \leq v)$ and $(y \star z \leq v)$ iff $(z \star x) \sqcup (z \star y) \leq v$.
(ii) $u \leq x \multimap (y \multimap z)$ iff $u \star x \leq y \multimap z$ iff $u \star x \star y \leq z$ iff $u \leq x \star y \multimap z$.
(iii) $\bot \leq x \multimap \bot \Rightarrow \bot \star x \leq \bot \Leftrightarrow x \leq \bot \multimap \bot$.
(iv) is left as an exercise. $\square$

♠ EXERCISE. Prove (iv) and the infinitary case under (i).

REMARK. (i) of the lemma holds because, in categorical terms, joins are colimits, and the functor $- \star z$, having a right adjoint, preserves colimits.

8.4.   PROPOSITION. *An alternative definition of IL-algebra is obtained replacing* (iii) *of definition* 8.2 *by* $z \star (x \sqcup y) = (z \star x) \sqcup (z \star y)$ *of the preceding lemma.*

PROOF. Assume (i), (ii), (iv) of the definition and $z \star (x \sqcup y) = (z \star x) \sqcup (z \star y)$. If $x \le x'$, then $x \sqcup x' = x'$. Hence $z \star (x \sqcup x') = (z \star x) \sqcup (z \star x') = z \star x'$, therefore $z \star x \le z \star x'$.

Also, assuming $x \le x'$: $z \le x' \multimap y$ iff $z \star x' \le y$, hence $z \star x \le y$, so $z \le x \multimap y$ and therefore $x' \multimap y \le x \multimap y$, etc. $\square$

8.5. Complete IL-algebras may be introduced in another way, namely as quantales.

DEFINITION. $\mathcal{X} \equiv (X, \sqcup, \star, \mathbf{1})$ is a *commutative quantale with unit* if $(X, \sqcup)$ is a complete lattice with infinitary join operator $\sqcup$, and $(X, \star, \mathbf{1})$ is a monoid, such that for all $x \in X$, $\{y_i : i \in I\} \subset X$

$$x \star \bigsqcup_{i \in I} y_i = \bigsqcup_{i \in I} (x \star y_i).$$

We shall use simply *quantale* for commutative quantale with unit. $\square$

PROPOSITION. *A quantale becomes a complete IL-algebra if we define*

$$\sqcap X := \bigsqcup \{y : \forall x \in X (y \le x)\}$$
$$x \multimap y := \bigsqcup \{z : x \star z \le y\}.$$

♠ EXERCISE. Prove this.

The proposition shows that complete IL-algebras and quantales amount to the same thing. But a quantale morphism, preserving $\sqcup, \star, \mathbf{1}$ is not necessarily an IL-morphism! On the other hand an IL-morphism preserving arbitrary joins is a quantale morphism.

8.6. DEFINITION. Let $\mathcal{X} \equiv (X, \sqcap, \sqcup, \bot, \multimap, \star, \mathbf{1})$ be an IL-algebra. $C : X \longrightarrow X$ is a *closure operation* on $\mathcal{X}$ if

  (i) $x \le Cx$,

  (ii) if $x \le y$ then $Cx \le Cy$,

  (iii) $CCx \le Cx$,

  (iv) $Cx \star Cy \le C(x \star y)$.

$x \in X$ is *C-closed* if $Cx = x$. $C(X)$ is the set of C-closed elements. $\square$

8.7. Lemma. $C(Cx \star Cy) = C(x \star y)$, and $C(X)$ *is closed under* $\sqcap$ *and* $\multimap$.

Proof. The proof of the first statement is left to the reader.

(ii) $C(Cx \multimap Cy) \le Cx \multimap Cy$ iff $C(Cx \multimap Cy) \star Cx \le Cy$. But $C(Cx \multimap Cy) \star Cx \le C((Cx \multimap Cy) \star Cx) \le CCy = Cy$ (using $(u \multimap v) \star u \le v$ and (i)–(iv) of the definition).

(iii) $Cx \sqcap Cy \le Cx, Cy$, hence $C(Cx \sqcap Cy) \le Cx, Cy$, therefore $C(Cx \sqcap Cy) \le Cx \sqcap Cy \le C(Cx \sqcap Cy)$. $\square$

♠ Exercise. Prove the first part.

8.8. Proposition. *Let* $\mathcal{X} \equiv (X, \sqcap, \sqcup, \bot, \multimap, \star, 1)$ *be an IL-algebra with closure operation* C, *put*

$$x \sqcup_c y := C(x \sqcup y), \quad x \star_c y := C(x \star y),$$

then

$$C(\mathcal{X}) \equiv (C(X), \sqcap, \sqcup_c, C\bot, \multimap, \star_c, C1)$$

is again an IL-algebra.

Proof. (i) $\sqcup_c$ is the join: $Cx, Cy \le C(x \sqcup y)$, and if $Cx \le Cz$, $Cy \le Cz$ then $Cx \sqcup Cy \le Cz$, hence $C(Cx \sqcup Cy) \le Cz$, so $Cx \sqcup_c Cy \le Cz$.

(ii) $\star_c$ is associative: $x \star_c (y \star_c z) = C(x \star C(y \star z)) = C(Cx \star C(y \star z))$ $= (8.7) \ C(x \star (y \star z)) = C((x \star y) \star z) = (8.7) \ C(C(x \star y) \star z) = (x \star_c y) \star_c z$.

(iii) $C1$ is the unit: $Cx \star_c C1 = C(Cx \star C1) = (8.7) \ C(x \star 1) = Cx$.

(iv) $\star$ distributes over $\sqcup$: $C(z \star C(x \sqcup y)) = C(Cz \star C(x \sqcup y)) =$ $(8.7) \ C(z \star (x \sqcup y)) = C((z \star x) \sqcup (z \star y)) \le C(C(z \star x) \sqcup C(z \star y)) =$ $(z \star_c x) \sqcup_c (z \star_c y)$.

The converse is obvious since $(z \star x) \sqcup (z \star y) \le z \star (x \sqcup y)$.

(v) The adjointness property holds: $C(Cx \star Cy) \le Cz$ iff $Cx \star Cy \le Cz$ iff $Cx \le Cy \multimap Cz$. $\square$

8.9. Proposition. *Let* $(M, \bullet, \varepsilon)$ *be a commutative monoid with unit* ε. *Put for* $X, Y \in P(M)$ *(the power set of* M*)*

$$X \star Y := \{x \bullet y : x \in X, \ y \in Y\}$$

$$X \multimap Y := \{z : \forall x \in X(z \bullet x \in Y)\}.$$

Then

$$\mathcal{P}(M) \equiv (P(M), \cap, \cup, \emptyset, \multimap, \star, \{\varepsilon\}),$$

where $\cap$, $\cup$ *are the usual set-theoretic operations, is an IL-algebra.*

Proof. Completely straightforward. $\square$

REMARK. $\mathcal{P}(M)$ satisfies a law not generally valid in IL-algebras: distributivity of the lattice operations.

The simplest way to see that distributivity of the lattice operations does not hold in general, is to verify that the sequent

$$A \sqcap (B \sqcup C) \Rightarrow (A \sqcap B) \sqcup (A \sqcap C)$$

is not derivable, since this means that the IL-algebra constructed from **ILL** or **CLL** by means of the Lindenbaum construction does not obey distributivity. Now it is easy to see that there is no cut-free proof possible for atomic A, B, C.

8.10. PROPOSITION. $C(\mathcal{P}(M))$, *the IL-algebra obtained from* $\mathcal{P}(M)$ *by a closure operation* C, *is complete as a lattice and satisfies*

$$Y \star_C C(\bigcup_{i \in I} X_i) = C(\bigcup_{i \in I} C(Y \star X_i)).$$

PROOF. We have to prove completeness, and for closed Y, X_i

$$C(Y \star C(\bigcup_{i \in I} X_i)) = C(\bigcup_{i \in I} C(Y \star X_i)).$$

(i) $X_i \subset \bigcup X_i$, so $X_i = CX_i \subset C(\bigcup X_i)$. If $Y \supset X_i$ for all i, then $Y \supset \bigcup X_i$, $CY = Y \supset C(\bigcup X_i)$, so $C(\bigcup X_i)$ is indeed the least upper bound of the X_i within the set of closed elements.

(ii) $Y \star C(\bigcup X_i) \supset Y \star X_i$, so $C(Y \star C(\bigcup X_i)) \supset C(Y \star X_i)$, hence $C(Y \star C(\bigcup X_i)) \supset \bigcup C(Y \star X_i)$, and therefore $C(Y \star C(\bigcup X_i)) \supset C(\bigcup C(Y \star X_i))$. Conversely, $Y \star \bigcup X_i = \bigcup (Y \star X_i)$, so $C(CY \star C(\bigcup X_i)) = (8.7) \ C(Y \star \bigcup X_i) = C(\bigcup (Y \star X_i)) \subset C(\bigcup C(Y \star X_i))$. $\square$

DEFINITION. An algebra of the form $C(\mathcal{P}(M))$ is called a *phase structure*. $\square$

8.11. DEFINITION. Let $\mathcal{X} \equiv (X, \sqcap, \sqcup, \perp, \multimap, \star, \mathbf{1})$ be an IL-algebra and let $M = (X, \star, \mathbf{1})$ be its monoid. Define on subsets $Y \subset X$ the operation C

$$C(Y) := \{z : \forall x (\forall y \in Y (y \leq x) \to z \leq x)\}.$$

We often write Y^C for $C(Y)$ ($C(Y)$ is the *order completion* or *Mac-Neille completion* of Y). $\square$

8.12. LEMMA. C *is a closure operation on* $\mathcal{P}(M)$.
PROOF. (i) and (ii) of the definition of closure operation are immediate. As to (iii), the property $CC(Y) \subset C(Y)$, we observe that $C(Y)$ is the collection of lower bounds of the collection of upper bounds Y^* of Y.

Now $Y^* \subset C(Y)^*$; for if x is an upper bound of Y, then each element y of $C(Y)$, being a lower bound of the upper bounds of Y is below x, so x is an upper bound of $C(Y)$. And since $Y^* \subset Y^{*C}$, the lower bounds of Y^{*C} are lower bounds of Y^*, so $C(Y) \supset CC(Y)$.

(iv) of the definition: let $x \star y \in X^C \star Y^C$, and let $\forall u \in X \star Y(u \le v)$. We have to show $x \star y \le v$. $\forall u \in X \star Y(u \le v)$ means that $\forall x' \in X \forall y' \in Y(x' \star y' \le v)$, so in particular $\forall x' \in X(x' \le y' \multimap v)$ for all $y' \in Y$; since $x \in X^C$, it follows that $\forall y' \in Y(x \le y' \multimap v)$, hence $\forall y' \in Y(x \star y' \le v)$, so $\forall y' \in Y(y' \le x \multimap v)$; and since $y \in Y^C$ it follows that $y \le x \multimap v$, so $x \star y \le v$. $\square$

8.13. PROPOSITION. *Let* $\mathcal{X} \equiv (X, \sqcap, \sqcup, \bot, \multimap, \star, 1)$ *be an IL-algebra and let* C *be defined as above. Then the map* h *defined by*

$$h(a) := \{a\}^C = \{z : z \le a\}$$

is an IL-embedding (i.e., an injective IL-morphism) which preserves existing arbitrary joins and meets.
PROOF. (i) h is injective. For if $h(a) = h(b)$, then, since $a \in h(a)$, also $a \in h(b)$, which implies $a \le b$; similarly $b \le a$, so $a = b$.

(ii) h preserves $\star$ since $\{a\}^C \star_C \{b\}^C = C(C\{a\} \star \{b\}) = C(\{a \star b\})$ by Lemma 8.7.

(iii) h preserves $\multimap$ by Lemma 8.7.

(iv) Let $\sqcap A$ exist. $h(\sqcap A) = \bigcap h[A] (= \bigcap\{h(a) : a \in A\})$ is proved as follows: for $a \in A$, $\sqcap A \le a$, so $h(\sqcap A) \subset h(a)$, so $h(\sqcap A) \subset \bigcap\{h(a) : a \in A\}$. Conversely, if $z \in h(a)$ for all $a \in A$, we have $\forall a \in A(z \le a)$, so $z \le \sqcap A$ and therefore $z \in h(\sqcap A)$; hence $\bigcap h[A] \subset h(\sqcap A)$.

(v) Let $\sqcup A$ exist. then if $x \in A$, $h(a) \subset h(\sqcup A)$, so $\bigcup h[A] \subset h(\sqcup A)$, and therefore $C(\bigcup h[A]) \subset h(\sqcup A)$ since $h(\sqcup A)$ is closed. Conversely, let $x \in h(\sqcup A)$, then $x \le \sqcup A$ so x is a lower bound of the upper bounds of A, i.e., $x \in C(A) \subset C(\bigcup h[A])$. $\square$

8.14. DEFINITION. (IL-model) An *IL-structure* consists of a domain D and a complete IL-algebra. An *IL-model* is an IL-structure with a valuation $[\![\,]\!]$ assigning a value $[\![P]\!]$ to each atomic sentence P in the language extended with constants for elements of the domain D.

[ ] is extended to arbitrary formulas by

$$
\begin{aligned}
[\mathbf{1}] &:= \mathbf{1} \\
[\bot] &:= \bot \\
[A \diamond B] &:= [A] \diamond [B] \text{ for } \diamond \in \{\star, \sqcup, \sqcap, \multimap\} \\
[\forall x\, A(x)] &:= \textstyle\bigsqcap_d [A(d)] \\
[\exists x\, A(x)] &:= \textstyle\bigsqcup_d [A(d)].
\end{aligned}
$$

[ ] is extended to multisets by

$$
\begin{aligned}
[\Lambda] &:= \mathbf{1} \\
[\Gamma, \Delta] &:= [\Gamma] \star [\Delta]
\end{aligned}
$$

A sequent $\Gamma \Rightarrow A$ is *valid* iff $[\Gamma] \leq [A]$.

An *ILZ-structure* (*CL-structure*) is a domain with a complete ILZ-algebra (CL-algebra). An *ILZ-model* (*CL-model*) is defined as an IL-model, but now we require in addition for [ ]

$$
[\mathbf{0}] := \mathbf{0},
$$

and we stipulate that $\Gamma \Rightarrow \Lambda$ is valid iff $\Gamma \Rightarrow \mathbf{0}$ is valid iff $[\Gamma] \leq [\mathbf{0}]$.
□

By a routine induction on the length of derivations one proves

THEOREM. *(Soundness) If $\vdash \Gamma \Rightarrow A$ in $\mathbf{ILL_q}$ then $\Gamma \Rightarrow A$ is valid in every IL-model. Similarly for $\mathbf{ILZ_q}(\mathbf{CLL_q})$ and ILZ-models (CL-models).*

We also have the converse:

8.15. THEOREM. *(Completeness) For a suitable IL-model, if $\Gamma \Rightarrow A$ is valid in the model, then $\Gamma \Rightarrow A$ is provable in $\mathbf{ILL_q}$. Similarly for ILZ-models (CL-models) and $\mathbf{ILZ_q}$ ($\mathbf{CLL_q}$).*

PROOF. Without loss of generality we may assume $\Gamma = \Lambda$. Consider the Lindenbaum algebra of $\mathbf{ILL_q}$, and construct a phase structure from this according to 8.11. Let the (countable) collection of variables be the domain, and choose a valuation on the prime formulas by

$$
[P] = C(P/\Leftrightarrow);
$$

here $P/\Leftrightarrow$ is the equivalence class corresponding to P in the Lindenbaum algebra. It is now readily seen from Lemma 8.13 that $[A] = C(A/\Leftrightarrow) = h(A/\Leftrightarrow)$ for all A. It follows that, if A is valid, i.e., $[\mathbf{1}] \subset [A]$, $\vdash \mathbf{1} \multimap A$, that is to say $\vdash A$.

The extension to $\mathbf{ILZ_q}$ is trivial; if $\mathcal{X} \equiv (X, \sqcap, \sqcup, \perp, \multimap, \star, \mathbf{1}, \mathbf{0})$ is an ILZ-algebra and C a closure operation, then $(CX, \sqcap, \sqcup_c, C\perp, \multimap, \star_c, C\mathbf{1}, C\mathbf{0})$ is an ILZ-algebra. More specifically, for the C defined in 8.11,

$$(P(X), \cap, \cup_c, C\{\emptyset\}, \multimap, \star_c, C\{\mathbf{1}\}, C\{\mathbf{0}\})$$

is an ILZ-algebra. For $\mathbf{CLL_q}$ we need only a slight additional consideration. The just mentioned ILZ-algebra is in fact a CL-algebra since we can prove

$$(Y \multimap C\{\mathbf{0}\}) \multimap C\{\mathbf{0}\} = CY$$

for all $Y \in P(X)$. To see this, let $v \in ((Y \multimap C\{\mathbf{0}\}) \multimap C\{\mathbf{0}\})$, i.e.,

$$(1) \qquad \forall u(\forall y \in Y(y \star u \leq \mathbf{0}) \rightarrow v \star u \leq \mathbf{0}).$$

We have to show

$$(2) \qquad \forall u(\forall y \in Y(y \leq u) \rightarrow v \leq u).$$

Apply (1) to $\sim w$ and then use $\sim\sim w = w$, then

$$\forall y \in Y(\sim w \star y \leq \mathbf{0}) \rightarrow v \star \sim w \leq \mathbf{0},$$

which is equivalent to

$$\forall y \in Y(y \leq w) \rightarrow v \leq w,$$

which is (2) for $w = u$; the converse is similar. As a result, the completeness proof automatically extends to $\mathbf{CLL_q}$. $\square$

8.16. DEFINITION. Let $\mathcal{X} \equiv (X, \sqcap, \sqcup, \perp, \multimap, \star, \mathbf{1})$ be an IL-algebra. $! : X \longrightarrow X$ is a *modality* over $\mathcal{X}$ iff

(i) $\forall x \in X(!x \leq x)$

(ii) $\forall x, y \in X(!y \leq x \Rightarrow !y \leq !x)$

(iii) $\mathbf{1} = !\top$, where $\top := \perp \multimap \perp$

(iv) $\forall x, y \in X(!x \star !y = !(x \sqcap y))$

We call $(X, \sqcap, \sqcup, \perp, \multimap, \star, \mathbf{1}, !)$ an *ILS-algebra* (IL-algebra with *storage*). An *ILS-morphism* is an IL-morphism which in addition preserves ! Similarly for *CLS-algebra* and *CLS-morphism*. $\square$

REMARK. Clause (ii) of the definition is in fact equivalent to the requirement of monotonicity: $x \leq y \rightarrow !x \leq !y$ and $!x \leq !!y$.

8.17. LEMMA.

(i) $!!x = !x$

(ii) $1 = !1$

PROOF
(i) $!!x \leq !x$; and since $!x \leq !x$, also $!x \leq !!x$ by (ii) of the definition.
(ii) $!1 \leq 1$ by (i) of the definition, and $!\top = 1 \leq 1 \Rightarrow !\top \leq !1$, so
$1 \leq !1$. $\square$

♠ EXERCISE. Show that for any ILS-algebra $\mathcal{X} \equiv (X, \sqcap, \sqcup, \bot, \multimap, \star, 1, !)$,
$!\mathcal{X} := \{!a : a \in X\}$ is a Heyting algebra (for suitably defined $\wedge, \vee, \bot, \top$).

8.18. THEOREM. *Let $\mathcal{X} \equiv (X, \sqcap, \sqcup, \bot, \multimap, \star, 1)$ be an IL-algebra.
Let $F \subset X$ be such that*

(a) $\sqcup\{x \in F : x \leq y\}$ *exists for all $y \in X$,*

(b) *F is closed under $\star$,*

(c) *$x \star x = x$ for $x \in F$,*

(d) *$1 \in F, \forall x \in F(x \leq 1)$.*

Then $!_F(a) = \sqcup\{x \in F : x \leq a\}$ defines a modality over $\mathcal{X}$.
PROOF. We write $!$ for $!_F$.
  (i) of the definition is immediate.
  (ii) Assume $\sqcup\{z \in F : z \leq y\} \leq x$. Then $\forall z \in F(z \leq y \rightarrow z \leq x)$,
hence $!y \leq !x$.
  (iii) is immediate from condition (d) in the theorem.
  (iv) $!x \leq x$, $!y \leq y$, $!x \star !y \leq !x$, $!y$ (since $!x \star !y \leq !x \star !\top = !x \star 1 = !x$ etc.), hence $!x \star !y \leq x \sqcap y$. Also $!x \star !y = \sqcup_{z,z' \in F}\{z \star z' : z \leq x, z' \leq y\}$ (by distributivity of $\sqcup$ over $\star$) $\leq \sqcup_{z,z' \in F}\{z \star z' : z \leq !x, z' \leq !y\}$
(since $z \in F$ and $z \leq u$ implies $z \leq !u$) $\leq \sqcup\{z \in F : z \leq !x \star !y\} = !(!x \star !y)$; hence $!x \star !y = !(!x \star !y) \leq !(x \sqcap y)$.
  On the other hand $!(x \sqcap y) = \sqcup\{z \in F : z \leq x \sqcap y\}$, and $z \leq x \sqcap y \Rightarrow z \leq x, z \leq y$, which in turn implies $z \leq !x, z \leq !y$, hence $z = z \star z \leq !x \star !y$. So $!(x \sqcap y) \leq !x \star !y$. $\square$

8.19. THEOREM. *Let $\mathcal{X} \equiv (X, \sqcap, \sqcup, \bot, \multimap, \star, 1, !)$ be an ILS-algebra,
$\mathcal{Y}$ a complete IL-algebra, $f : \mathcal{X} \longrightarrow \mathcal{Y}$ an IL-embedding. Then $\mathcal{Y}$
can be extended to an ILS-algebra $\mathcal{Y}^*$ with operator $!^*$ such that f
becomes an ILS-embedding, i.e.,*

$$f(!x) = !^* f(x).$$

PROOF. Define
$$F := \{f(!x) : x \in X\},$$

then F is easily seen to satisfy the clauses (b)–(d) of the preceding theorem; and (a) is satisfied since $\mathcal{Y}$ is complete. So we can put

$$!^*a := !_F(a) = \bigsqcup\{x \in F : x \le a\}.$$

Now f is order preserving, so $f(!x) \le f(x)$, so $f(!x) \le !_F f(x) = \bigsqcup\{y \in F : y \le f(x)\}$. On the other hand, if $y \le f(x)$ for $y \in F$, then $y = f(!z) \le f(x)$ for some $z \in X$; f is an embedding preserving order, so $!z \le x$, hence $!z \le !x$; therefore $y = f(!z) \le f(!x)$ and so $!^*f(x) \le f(!x)$. $\square$

8.20. DEFINITION. The definition of valuation $[\![\ ]\!]$ in an ILS-algebra is the same as for a valuation in an IL-algebra, the definition is extended with a clause
$$[\![!A]\!] := ![\![A]\!].$$

ILS-structure and *ILS-model* are defined just as IL-structure, IL-model respectively. $\square$

8.21. THEOREM. *(Soundness) If* **ILL** $\vdash \Gamma \Rightarrow A$, *then* $\Gamma \Rightarrow A$ *holds for every valuation in every ILS-algebra.*
PROOF. Mainly routine. Let us consider the rule R! and assume $!\Gamma \Rightarrow A$ to be valid, then for $\Gamma = \{B_1, \ldots, B_n\}$ we have

$$![\![B_1]\!] \star \cdots \star ![\![B_n]\!] \le [\![A]\!].$$

Since $![\![B_1]\!] \star \cdots \star ![\![B_n]\!] = !([\![B_1]\!] \sqcap \cdots \sqcap [\![B_n]\!])$, this is a statement of the form $!x \le [\![A]\!]$, but then $!x \le ![\![A]\!] = [\![!A]\!]$, etc. $\square$

8.22. THEOREM. *(Completeness) There is a phase structure with modality such that if* $\Gamma \Rightarrow A$ *is valid in this structure, then* **ILL** $\vdash \Gamma \Rightarrow A$, *and similarly for* **ILZ, CLL.**

REMARK. Ono 1990b shows in addition that the existing proofs in the literature are all in essence the same; moreover in Proposition 8.13 h is an isomorphism if $\mathcal{X}$ is complete, and ! in an ILS-algebra is always representable as $!_F$ for an F satisfying the conditions in Theorem 8.18.

9 Combinatorial linear logic

9.1. In this chapter we describe versions of **ILL$_0$**, **ILL$_e$**, **CLL$_0$** and **CLL$_e$** inspired by the connections between linear logic and a special kind of categories. Sources are MacLane 1971, Lafont 1988a and Martí-Oliet & Meseguer 1990.

From a purely logical point of view, combinatorial (linear) logic may be described as a system for deriving sequents of the form $A \Rightarrow B$, i.e. with a single formula in the succedent as well as the antecedent; the system consists of some axiomatic sequents plus rules for deriving new sequents from old ones, e.g.,

$$\frac{A \Rightarrow B \qquad B \Rightarrow C}{A \Rightarrow C}$$

However, to bring out the connection with category theory, we shall introduce certain terms for describing the deductions in the system, just as we did in the introduction for the natural deduction version of conjunction logic. That is, nodes in the proof tree are now labeled with expressions

$$\tau : A \Rightarrow B$$

where τ is a term in a suitable term calculus, as before. The τ at the bottom node encodes in fact the whole tree.

9.2. Definition of C-ILL$_0$

We list the axioms and rules.

Sequential composition:

$$\frac{\phi : A \Rightarrow B \qquad \psi : B \Rightarrow C}{\psi \circ \phi : A \Rightarrow C} \qquad\qquad \mathrm{id}_A : A \Rightarrow A$$

Parallel composition:

$$\frac{\phi : A \Rightarrow B \qquad \psi : C \Rightarrow D}{\phi \star \psi : A \star C \Rightarrow B \star D} \qquad\qquad 1 : \mathbf{1} \Rightarrow \mathbf{1}$$

Adjointness of $\multimap$ and $\star$:

$$\frac{\phi : A \star B \Rightarrow C}{\mathrm{cur}_{A,B,C}(\phi) : A \Rightarrow B \multimap C} \qquad\qquad \mathrm{ev}_{A,B} : (A \multimap B) \star A \Rightarrow B$$

Symmetry, associativity and unit:

$$\gamma_{A,B} : A \star B \Rightarrow B \star A$$

$$\alpha_{A,B,C} : A \star (B \star C) \Rightarrow (A \star B) \star C \quad \alpha_{A,B,C}^{-1} : (A \star B) \star C \Rightarrow A \star (B \star C)$$

$$\lambda_A : \mathbf{1} \star A \Rightarrow A \qquad \lambda_A^{-1} : A \Rightarrow \mathbf{1} \star A$$

Products and coproducts:

$$\frac{\phi : A \Rightarrow B \qquad \psi : A \Rightarrow C}{\langle \phi, \psi \rangle : A \Rightarrow B \sqcap C} \qquad\qquad \pi_{A,B,i} : A_0 \sqcap A_1 \Rightarrow A_i \ (i \in \{0,1\})$$

$$\top_A : A \Rightarrow \top$$

$$\frac{\phi : B \Rightarrow A \qquad \psi : C \Rightarrow A}{[\phi, \psi] : B \sqcup C \Rightarrow A} \qquad\qquad \kappa_{A,B,i} : A_i \Rightarrow A_0 \sqcup A_1 \ (i \in \{0,1\})$$

$$\perp_A : \perp \Rightarrow A$$

Usually we shall drop the formula subscripts in id_A, $\mathrm{cur}_{A,B,C}$, $\mathrm{ev}_{A,B}$, $\gamma_{A,B}$, $\alpha_{A,B,C}$, $\alpha_{A,B,C}^{-1}$, λ_A, λ_A^{-1}, $\top_A$, $\perp_A$, $\pi_{A,B,i}$, $\kappa_{A,B,i}$ (so we write π_0, π_1, κ_0, κ_1 etc.)

9.3. THEOREM. N-$\mathbf{ILL}_0$ and C-$\mathbf{ILL}_0$ are equivalent in the following sense: if $\phi : A \Rightarrow B$ is derivable in C-$\mathbf{ILL}_0$, there is a deduction of $A \vdash B$ in N-$\mathbf{ILL}_0$; and if $A_1, \ldots, A_n \vdash B$ in N-$\mathbf{ILL}_0$, there is a ϕ such that $\phi : A_1 \star \cdots \star A_n \Rightarrow B$ is derivable in C-$\mathbf{ILL}_0$.

PROOF. Straightforward in both directions, by induction on the length of derivations. For example, consider a derivation in N-$\mathbf{ILL}_0$ ending with

$$\frac{\Gamma, A \vdash C \qquad \Gamma, B \vdash C \qquad \Delta \vdash A \sqcup B}{\Gamma, \Delta \vdash C}$$

Let D, D' be the tensor products (obtained by association to the left) of the elements of Γ and Δ respectively. By induction hypothesis we have

$$\phi : A \star D \Rightarrow C \qquad \phi' : B \star D \Rightarrow C \qquad \psi : D' \Rightarrow A \sqcup B$$

Then

$$\mathrm{cur}(\phi) : A \Rightarrow D \multimap C, \qquad \mathrm{cur}(\phi) : B \Rightarrow D \multimap C,$$

hence

$$[\mathrm{cur}(\phi), \mathrm{cur}(\phi')] : A \sqcup B \Rightarrow D \multimap C,$$

and so

$$[\mathrm{cur}(\phi), \mathrm{cur}(\phi')] \circ \psi : D' \Rightarrow D \multimap C$$

$$([\mathrm{cur}(\phi), \mathrm{cur}(\phi')] \circ \psi) \star \mathrm{id}_D : D' \star D \Rightarrow (D \multimap C) \star D,$$

$$\xi' \equiv \mathrm{ev} \circ (([\mathrm{cur}(\phi), \mathrm{cur}(\phi')] \circ \psi) \star \mathrm{id}_D) : D' \star D \Rightarrow C.$$

We can then find a term

$$\xi \equiv \xi' \circ \xi'' : D'' \Rightarrow C$$

where D'' is the tensor product representing the multiset Γ, Δ and $\xi'' : D'' \longrightarrow D' \star D$ is an isomorphism composed from (components of) $\alpha, \alpha^{-1}, \gamma$. $\square$

The proof is also easily given relative to the Gentzen-type sequent calculus $\mathbf{ILL}_0$.

♠ EXERCISE. Complete the proof.

9.4. C-$\mathbf{ILL}_0$ and intuitionistic linear categories

The calculus C-$\mathbf{ILL}_0$ can be made into a category with extra structure as follows. The formulas correspond to objects, and the proof terms (*combinators*) $\phi : A \Rightarrow B$ to arrows from A to B, i.e., in categorical notation $\phi : A \longrightarrow B$. id_A represents the identity on A, $\circ$ composition etc. However, we have to identify certain terms denoting arrows in order to make this into a category; thus the associativity of composition in a category requires

$$\phi \circ (\psi \circ \chi) = (\phi \circ \psi) \circ \chi,$$

and since id_A is an identity arrow, we must have

$$\mathrm{id}_A \circ \phi = \phi, \qquad \phi \circ \mathrm{id}_B = \phi \text{ for } \phi : A \longrightarrow B.$$

TABLE 10
Equations for intuitionistic linear categories

$\phi,\ \psi,\ \chi$ range over arrows; $\mathrm{id}_A : A \longrightarrow A$ is the identity; $\alpha,\ \gamma,\ \lambda$ are natural transformations with $\alpha_{A,B,C} : A \star (B \star C) \longrightarrow (A \star B) \star C$, $\gamma_{A,B} : A \star B \longrightarrow B \star A$, $\lambda_A : \mathbf{1} \star A \longrightarrow A$. If $\phi : A \star B \longrightarrow C$ then $\mathrm{cur}(\phi) : A \longrightarrow (B \multimap C)$.

Category axioms:

$$(\phi \circ \psi) \circ \chi = \phi \circ (\psi \circ \chi), \quad \mathrm{id} \circ \phi = \phi, \quad \phi \circ \mathrm{id} = \phi$$

Functorial character of $\star$:

$$(\phi \circ \phi') \star (\psi \circ \psi') = (\phi \star \psi) \circ (\phi' \star \psi'), \quad \mathrm{id} \star \mathrm{id} = \mathrm{id} \quad 1 = \mathrm{id}_{\mathbf{1}}$$

$\alpha,\ \gamma,\ \lambda$ are natural isomorphisms with inverses:

$$((\phi \star \psi) \star \chi) \circ \alpha = \alpha \circ (\phi \star (\psi \star \chi)), \quad \alpha \circ \alpha^{-1} = \mathrm{id}, \quad \alpha^{-1} \circ \alpha \overset{.}{=} \mathrm{id}$$
$$\gamma \circ (\phi \star \psi) = (\psi \star \phi) \circ \gamma, \quad \gamma \circ \gamma = \mathrm{id}$$
$$\lambda \circ (1 \star \phi) = \phi \circ \lambda, \quad \lambda \circ \lambda^{-1} = \mathrm{id} \quad \lambda^{-1} \circ \lambda = \mathrm{id}$$

Coherence conditions:

$$(\alpha \star \mathrm{id}) \circ \alpha \circ (\mathrm{id} \star \alpha) =$$
$$\alpha \circ \alpha : A \star (B \star (C \star D)) \to ((A \star B) \star C) \star D,$$
$$(\lambda \star \mathrm{id}) \circ \alpha = \lambda : \mathbf{1} \star (B \star C) \longrightarrow B \star C, \quad \lambda_{\mathbf{1}} \circ \gamma = \lambda_{\mathbf{1}}$$
$$\alpha \circ \alpha = (\gamma \star \mathrm{id}) \circ \alpha \circ (\mathrm{id} \star \gamma) : A \star (B \star C) \longrightarrow (C \star A) \star B$$

Equations for $x \multimap -$ as adjoint of $- \star x$:

$$\mathrm{cur}(\phi) \circ \psi = \mathrm{cur}(\phi \circ (\psi \star \mathrm{id}))$$
$$\mathrm{ev} \circ (\mathrm{cur}(\phi) \star \psi) = \phi \circ (\mathrm{id} \star \psi)$$
$$\mathrm{cur}(\mathrm{ev}) = \mathrm{id} : (A \multimap B) \longrightarrow (A \multimap B)$$

Equations for products:

$$\pi_1 \circ \langle \phi, \psi \rangle = \phi, \quad \pi_2 \circ \langle \phi, \psi \rangle = \psi, \quad \langle \phi, \psi \rangle \circ \chi = \langle \phi \circ \chi, \psi \circ \chi \rangle$$
$$\langle \pi_1, \pi_2 \rangle = \mathrm{id}_{A \sqcap B}$$
$$\top_A \circ \phi = \top_B : B \longrightarrow \top \text{ if } \phi : B \longrightarrow A, \quad \top_\top = \mathrm{id}_\top$$

Equations for coproducts:

$$[\phi, \psi] \circ \kappa_1 = \phi, \quad [\phi, \psi] \circ \kappa_2 = \psi, \quad \chi \circ [\phi, \psi] = [\chi \circ \phi, \chi \circ \psi]$$
$$[\kappa_1, \kappa_2] = \mathrm{id}_{A \sqcup B}$$
$$\phi \circ \bot_A = \bot_B : \bot \longrightarrow B \text{ if } \phi : A \longrightarrow B, \quad \bot_\bot = \mathrm{id}_\bot$$

The full set of equations is listed in Table 10; formula subscripts have been mostly dropped in order not to encumber the notation too much. We briefly discuss the significance of the various groups of axioms.

(i) As already stated, the category axioms make the graph into a category $\mathcal{C}$.

(ii) $-\star-$ is a functor from $\mathcal{C} \times \mathcal{C}$ into $\mathcal{C}$.

(iii) α, γ, λ are natural isomorphisms with inverses α^{-1}, γ^{-1}, λ^{-1} respectively; α expresses associativity of $\star$, γ expresses symmetry, and λ expresses that $\mathbf{1}$ behaves as a "neutral element" w.r.t. $\star$. For example, the first equation for α says that all squares of the following form commute:

$$
\begin{array}{ccc}
A \star (B \star C) & \xrightarrow{\alpha_{A,B,C}} & (A \star B) \star C \\
\downarrow{\phi \star (\psi \star \chi)} & \circledast & \downarrow{(\phi \star \psi) \star \chi} \\
A' \star (B' \star C') & \xrightarrow{\alpha_{A',B',C'}} & (A' \star B') \star C'
\end{array}
$$

(iv) The coherence equations express the commutativity of the pentagon diagram (the *pentagon condition*):

$$
\begin{array}{ccccc}
A(B(CD)) & \xrightarrow{\alpha} & (AB)(CD) & \xrightarrow{\alpha} & ((AB)C)D \\
\downarrow{\text{id} \star \alpha} & & \circledast & & \uparrow{\alpha \star \text{id}} \\
A((BC)D) & & \xrightarrow{\alpha} & & (A(BC))D
\end{array}
$$

of the triangle diagram (the *triangle condition*):

$$
\begin{array}{ccc}
\mathbf{1}(BC) & \xrightarrow{\alpha} & (\mathbf{1}B)C \\
\downarrow{\lambda} & \circledast & \downarrow{\lambda \star \text{id}} \\
BC & = & BC
\end{array}
$$

and of the hexagon diagram (the *hexagon condition*):

$$
\begin{array}{ccccc}
A(BC) & \xrightarrow{\text{id} \star \gamma} & A(CB) & \xrightarrow{\alpha} & (AC)B \\
\downarrow{\alpha} & & \circledast & & \downarrow{\gamma \star \text{id}} \\
(AB)C & \xrightarrow{\gamma} & C(AB) & \xrightarrow{\alpha} & (CA)B
\end{array}
$$

Together with $\gamma_{1,1} = \text{id}_{1\star 1}$, this means that the resulting category is *symmetric monoidal.*

(v) The equations for cur and ev express that $A \multimap -$ is left adjoint to $- \star A$. In other words, to each arrow $\phi : A \star B \longrightarrow C$ there is a *unique arrow* $\text{cur}(\phi) : A \longrightarrow (B \multimap C)$ such that

$$
\begin{array}{ccc}
A \star B & = & A \star B \\
\Big\downarrow{\scriptstyle \text{cur}(\phi)\,\star\,\text{id}} & \circledast & \Big\downarrow{\scriptstyle \phi} \\
(B \multimap C) \star B & \xrightarrow{\;\;\text{ev}\;\;} & C
\end{array}
$$

commutes. $\text{ev} \circ (\text{cur}(\phi) \star \text{id}) = \phi$ and $\text{cur}(\text{ev} \circ (\psi \star \text{id})) = \psi$ express commutativity of the triangle and uniqueness of $\text{cur}(\phi)$ respectively. The equations in Table 10 are easily seen to be equivalent.

♠ EXERCISE. Prove the equivalence.

(vi) The product equations tell us that $A \sqcap B$ is the categorical product of A and B, with projections π_0 and π_1, and that $\top$ is the terminal object.

Similarly, the coproduct equations tell us that $A \sqcup B$ is the coproduct of A and B, and that $\bot$ is the initial object.

Note that the axioms for $\top_A$ may be replaced by

$$\top_A = \phi \text{ for any } \phi : A \longrightarrow \top,$$

and the axioms for $\bot_A$ by

$$\bot_A = \phi \text{ for any } \phi : \bot \longrightarrow A.$$

9.5. DEFINITION. A category $\mathcal{C}$ with $\star, 1, \alpha, \gamma, \lambda$ as above is a *symmetric monoidal category* (SMC), a category with $\star, 1, \alpha, \gamma, \lambda, \text{cur}, \text{ev}$ is a *symmetric monoidal closed category* (SMCC), a category with $\star, 1, \alpha, \gamma, \lambda, \text{cur}, \text{ev}, \pi_0, \pi_1, \langle \rangle, \kappa_0, \kappa_1, [\,], \top, \bot$ is called an *intuitionistic linear category* (ILC).

REMARK. In the absence of γ, i.e., for monoidal and monoidal closed categories, we need besides λ also a natural transformation ρ such that

$$\text{id} \star \rho_C = \rho_{B\star C} \circ \alpha_{B,C,1}, \quad \lambda_1 = \rho_1.$$

In the presence of γ, ρ is definable by $\rho_A = \lambda_A \circ \gamma$.

9.6. EXAMPLES

(i) The category obtained from the formalism C-**ILL**$_0$ itself is a free ILC constructed over a discrete graph (with the prime formulas as nodes and without axioms). More generally we may construct a free ILC$(\mathcal{G})$ over an arbitrary directed graph.

(ii) IL-algebras are a degenerate example, with a partially ordered set as underlying category. Here all the identifications imposed by the equations in the table trivialize, since there is always at most one arrow between any two objects in a poset.

(iii) The category **Set*** of pointed sets, with as objects sets X which all contain a designated element $*_X$ (usually simply $*$), and as arrows all set theoretic mappings f from X to Y with $f(*_X) = *_Y$.

$X \star Y$ is defined as $(X \times Y)/ \sim$, where $\sim$ is an equivalence relation on $X \times Y$, identifying all pairs of the form $(*_X, z)$ or $(z, *_Y)$ into $*_{X \star Y}$, and nothing else. The tensor unit is the set $\{*, a\}$ with a some element distinct from $*$.

The categorical product $X \sqcap Y$ is simply the cartesian product with $*_{X \sqcap Y} = (*_X, *_Y)$, the categorical coproduct $X \sqcup Y$ is the disjoint sum of X and Y with $*_{X \sqcup Y} = *_X = *_Y$ identified. $\top$ and **0** are both $\{*\}$.

(iv) Every cartesian closed category (CCC) is a degenerate example of a SMCC, where $\star$ and $\sqcap$, **1** and $\top$, **0** and $\perp$ are pairwise identified. So the equations for a CCC consist of the category axioms, equations for products and coproducts, and the equations for $x \multimap -$ as adjoint to $- \sqcap x$ (i.e., $\rightarrow$, $\wedge$ replace $\multimap$ and $\star$ respectively in the equations for $\multimap$). Alternatively, we may choose the slightly different versions (more directly translating the categorical definitions) which were mentioned in our discussion above. We have listed the equations for a CCC in Table 11. (In the literature coproducts are not required for a CCC, but here we use "CCC" indiscriminately both for CCC's in the narrow sense and for CCC's with coproducts.)

9.7. Axioms and rules for !

The exponential ! may be added to C-**ILL**$_0$, producing a system C-**ILL**$_e$. We have to add

$$\frac{\phi : A \Rightarrow B}{!\phi : !A \Rightarrow !B}, \qquad \mathbf{s}_A : !A \Rightarrow !!A, \qquad \mathbf{r}_A : !A \Rightarrow A,$$

$$\mathbf{t} : !\top \Rightarrow \mathbf{1},$$

$$\mathbf{p}_{A,B} : !(A \sqcap B) \Rightarrow !A \star !B, \qquad \mathbf{p}_{A,B}^{-1} : !A \star !B \Rightarrow !(A \sqcap B).$$

TABLE 11
Equations for CCC's

Category axioms:

ass $: (\phi \circ \psi) \circ \chi = \phi \circ (\psi \circ \chi)$,   idl $:$ id $\circ \phi = \phi$, idr $: \phi \circ$ id $= \phi$

Equations for products:

$$\text{prl} : \pi_0 \circ \langle \phi, \psi \rangle = \phi, \quad \text{prr} : \pi_1 \circ \langle \phi, \psi \rangle = \psi$$
$$\text{pair} : \langle \phi, \psi \rangle \circ \chi = \langle \phi \circ \chi, \psi \circ \chi \rangle, \quad \text{pairid} : \langle \pi_0, \pi_1 \rangle = \text{id}_{A \wedge B}$$
$$\text{ter} : \top_A \circ \phi = \top_B : B \longrightarrow \top \text{ if } \phi : B \longrightarrow A, \quad \text{terid} : \top_\top = \text{id}_\top$$

Equations for coproducts:

$$\text{inl} : [\phi, \psi] \circ \kappa_0 = \phi, \quad \text{inr} : [\phi, \psi] \circ \kappa_1 = \psi,$$
$$\text{uni} : [\phi \circ \psi] = [\chi \circ \phi, \chi \circ \psi], \quad \text{unid} : [\kappa_0, \kappa_1] = \text{id}_{A \vee B}$$
$$\text{init} : \phi \circ \bot_A = \bot_B : \bot \longrightarrow B \text{ if } \phi : A \longrightarrow B, \quad \text{initid} : \bot_\bot = \text{id}_\bot$$

Equations for $x \to -$ as adjoint to $- \wedge x$:

$$\text{cur} : \text{cur}(\phi) \circ \psi = \text{cur}(\phi \circ \langle \psi \circ \pi_0, \pi_1 \rangle)$$
$$\text{ev} : \text{ev} \circ (\text{cur}(\phi) \wedge \psi) = \phi \circ (\text{id} \wedge \psi)$$
$$\text{curid} : \text{cur}(\text{ev}) = \text{id} : (A \to B) \longrightarrow (A \to B)$$

VARIANT

Category axioms

as above

Equations for products:

$$\pi_0 \circ \langle \phi, \psi \rangle = \phi, \quad \pi_1 \circ \langle \phi, \psi \rangle = \psi, \quad \langle \pi_0 \circ \xi, \pi_1 \circ \xi \rangle = \xi$$
$$\top_A = \phi \text{ for all } \phi : A \longrightarrow \top$$

Equations for coproducts:

$$[\phi, \psi] \circ \kappa_0 = \phi, \quad [\phi, \psi] \circ \kappa_1 = \psi, \quad [\xi \circ \kappa_0, \xi \circ \kappa_1] = \xi$$
$$\bot_A = \phi \text{ for all } \phi : \bot \longrightarrow A$$

Equations for the exponential:

$$\text{ev} \circ \langle \text{cur}(\phi), \text{id} \rangle = \phi, \quad \text{cur}(\text{ev} \circ \langle \phi, \text{id} \rangle) = \phi$$

PROPOSITION. N-**ILL**$_e$ and C-**ILL**$_e$ are equivalent in the sense of Theorem 9.3.

PROOF. We check that the list above suffices. Thinning is justified by:

$$\frac{\dfrac{\mathbf{1}\star B \Rightarrow B \quad B \Rightarrow C}{\mathbf{1}\star B \Rightarrow C} \quad \dfrac{!\mathsf{T} \Rightarrow \mathbf{1} \quad B \Rightarrow B}{!\mathsf{T}\star B \Rightarrow \mathbf{1}\star B}}{!\mathsf{T}\star B \Rightarrow C} \qquad \frac{\dfrac{A \Rightarrow \mathsf{T}}{!A \Rightarrow !\mathsf{T}} \quad B \Rightarrow B}{!A \star B \Rightarrow !\mathsf{T}\star B}$$

$$!A \star B \Rightarrow C$$

We derive first

$$\frac{!C \Rightarrow !!C \quad \dfrac{\dfrac{C \Rightarrow C \quad C \Rightarrow C}{!C \Rightarrow C \quad C \Rightarrow C \sqcap C}}{!!C \Rightarrow !(C \sqcap C)}}{!C \Rightarrow !(C \sqcap C)} \qquad !(C \sqcap C) \Rightarrow !C \star !C}{!C \Rightarrow !C \star !C}$$

and then contraction is justified by

$$\frac{A \star (!C \star !C) \Rightarrow D \quad \dfrac{A \Rightarrow A \quad !C \Rightarrow !C \star !C}{A \star !C \Rightarrow A \star (!C \star !C)}}{A \star !C \Rightarrow D}$$

etc. We leave the rest of the verification to the reader.

How do we extend the equations of Table 10 to cover the operator *of course* as well? As we shall see, the answer is not unique. One has to find a "natural" notion of category with additional structure in which a functor ! exists, and arrows s_A, r_A, t, $p_{A,B}$, $p_{A,B}^{-1}$ which satisfy suitable equations.

A more or less "minimal" solution is given in Table 12.

9.8.   COMMENT. The naturality of s and r means that the following two squares commute:

$$
\begin{array}{ccc}
!A & \xrightarrow{\;\;s_A\;\;} & !!A \\
\Big\downarrow{\scriptstyle !\phi} \quad \circledast & & \Big\downarrow{\scriptstyle !!\phi} \\
!B & \xrightarrow{\;\;s_B\;\;} & !!B
\end{array}
\qquad
\begin{array}{ccc}
!A & \xrightarrow{\;\;r_A\;\;} & A \\
\Big\downarrow{\scriptstyle !\phi} \quad \circledast & & \Big\downarrow{\scriptstyle \phi} \\
!B & \xrightarrow{\;\;r_B\;\;} & B
\end{array}
$$

TABLE 12

Equations for the exponentials in an intuitionistic linear category

! is functorial:
$$!(\phi \circ \psi) = !\phi \circ !\psi, \quad !(\mathrm{id}) = \mathrm{id}$$

$\mathbf{s} : ! \overset{.}{\to} !!$ is natural: for $\phi : A \to B$

$$(!!\phi) \circ \mathbf{s}_A = \mathbf{s}_B \circ (!\phi)$$

$\mathbf{r} : ! \overset{.}{\to} \mathrm{id}$ is natural: for $\phi : A \longrightarrow B$

$$\phi \circ \mathbf{r}_A = \mathbf{r}_B \circ (!\phi)$$

$(!, \mathbf{s}, \mathbf{r})$ is a comonad:

$$!\mathbf{s}_A \circ \mathbf{s}_A = \mathbf{s}_{!A} \circ \mathbf{s}_A, \quad \mathrm{id}_{!A} = \mathbf{r}_{!A} \circ \mathbf{s}_A = (!\mathbf{r}_A) \circ \mathbf{s}_A$$

$\mathbf{t}$ and $\mathbf{p}$ are iso:
$$\mathbf{t} \circ \mathbf{t}^{-1} = \mathrm{id}, \quad \mathbf{t}^{-1} \circ \mathbf{t} = \mathrm{id},$$
$$\mathbf{p} \circ \mathbf{p}^{-1} = \mathrm{id}, \quad \mathbf{p}^{-1} \circ \mathbf{p} = \mathrm{id}$$

The comonad equations state the commutativity of the following diagrams:

(If we take for ! an arbitrary functor $L : \mathcal{C} \longrightarrow \mathcal{C}$, for s, r natural transformations $\delta : L \overset{.}{\longrightarrow} L^2$, $\epsilon : L \overset{.}{\longrightarrow} I$, then this is just the definition of a *comonad* in category theory).

DEFINITION. $\mathcal{C} \equiv (\mathcal{C}, \star, \mathbf{1}, \alpha, \gamma, \lambda, !, \mathbf{s}, \mathbf{r}, \mathbf{t}, \mathbf{p})$ is an *ILC with storage (ILCS)* if $(!, \mathbf{s}, \mathbf{r})$ is a comonad and $\mathbf{t}, \mathbf{p}$ are natural isomorphisms, as in Table 12. □

REMARK. With the comonad $M = (L, \delta, \varepsilon)$ in $\mathcal{C}$ we can associate the co-Kleisli category $\mathcal{C}_M$, with as objects the objects of $\mathcal{C}$, and as arrows $f \in \mathcal{C}_M(x, y)$ the arrows $f \in \mathcal{C}(Lx, y)$ with as composition of

$f : LX \longrightarrow y, \; g : Ly \longrightarrow z$

$$g \circ_M f := g \circ Lf \circ \delta_x$$

(cf. MacLane 1971, VI 5 for the dual Kleisli category).

♠ EXERCISE. Show that $\mathcal{C}_M$ is again a category, and show for $M \equiv (!, \mathbf{s}, \mathbf{r})$ as above that M is cartesian closed.

There remains the question what are the natural categorical counterparts to $\mathbf{CLL}_0$ and $\mathbf{CLL}_e$. The most promising proposal so far is due to Martí-Oliet and Meseguer (1990), simplifying an earlier proposal by Seely (1989) using *-autonomous categories. We shall not pursue the theory of these "classical linear categories" here, but give the definition only.

9.9. DEFINITION. Let $\mathcal{C} \equiv (\mathcal{C}, \star, \mathbf{1}, \alpha, \gamma, \lambda)$ be an SMC; let θ be the natural transformation given by

$$\theta_{A,B} := \mathrm{cur}(\mathrm{ev}_{A,B} \circ \gamma_{A, A \multimap B}) : A \longrightarrow (A \multimap B) \multimap B.$$

$\mathbf{0}$ is said to be a *dualizing object* if ν given by

$$\nu_A := \theta_{A,\mathbf{0}}$$

is a natural isomorphism. A *classical linear category*, or *linear category* for short (a CLC) is an intuitionistic linear category with dualizing object. A *CLC with storage (a CLCS)* is defined similar to an ILC with storage. □

REMARK. It should be pointed out that there is a good deal of freedom in associating a category with a term system for a logic; the identifications one makes between terms are dictated by the wish to arrive at a manageable and interesting class of categories as models.

9.10. EXAMPLES

(i) Let K be a field, Vec_K the category of finite-dimensional vector spaces over K with linear maps. Taking tensor product of vector spaces for $\star$, and interpreting $V \multimap W$ as the vector space of linear maps from V into W, we obtain a CLC with dualizing object K (treated as a one-dimensional vector space).

(ii) Another important example is the category Lin introduced in the next chapter.

♠ EXERCISE. Show in detail that Vec_K is a CLC.

10   Girard domains

10.1.   We now turn to an interesting type-theoretic model of **CLL**, the model of the *Girard domains*, also called *coherence spaces* (Girard 1986, 1987). Our exposition is based in part on Lafont 1988c. It is a type-theoretic model, not a logical model, since all types are inhabited, and moreover $\perp$ and **0** obtain the same interpretation as $\top$ and **1** respectively.

The treatment below is self-contained, modulo some notions of category theory. However, some background in domain theory is convenient; see e.g., Scott 1982 (for Scott domains) and Jung 1989.

10.2.   DEFINITION.   A *web* $\mathbf{A} \equiv (A, \sim_A)$ is a pair consisting of a set A and a symmetric and reflexive relation $\sim_A$. (In this chapter we use boldface capitals for webs).

$$\mathrm{Coh}(\mathbf{A}) := \{x \subset A : \forall \alpha, \beta \in x(\alpha \sim_A \beta)\}$$

is the set of *coherent subsets* of **A**. The collection of coherent subsets of a web, ordered under inclusion, is called a *Girard domain*. Usually we shall write $\mathcal{A}, \mathcal{B}, \ldots$ etc. for $(\mathrm{Coh}(A, \sim_A), \subset)$, $(\mathrm{Coh}(B, \sim_B), \subset)$, $\ldots$ etc. The elements of A are called *tokens* or *atoms* of the domain $\mathcal{A}$.

Fincoh(**A**) or $\mathcal{A}_{\mathrm{fin}}$ consists of the finite coherent subsets of **A**. $\square$

The tokens of a Girard domain represent atomic bits of information; a coherent set is a consistent piece of information. Coherence of tokens means that the tokens may be regarded as bits of information concerning the same object. The order of information is reflected by inclusion: $a \subset b$ means that b represents more information than a.

10.3. PROPOSITION. *For any Girard domain $\mathcal{A}$:*

(i) $\mathcal{A}$ *contains all singletons* $\{\alpha\} \subset \mathcal{A}$,

(ii) $a \in \mathcal{A}$, $b \subset a \Rightarrow b \in \mathcal{A}$,

(iii) $B \subset \mathcal{A}$, $\forall c, c' \in B(c \cup c' \in \mathcal{A}) \Rightarrow \bigcup B \in \mathcal{A}$,

(iv) $\emptyset \in \mathcal{A}$,

(v) $\mathcal{A}$ *closed w.r.t. directed unions (i.e., directed w.r.t. $\subset$),*

(vi) $\mathcal{A}$ *closed under inhabited intersections, i.e.,*
$a_i \in \mathcal{A}$ *for all* $i \in I$, I *inhabited* $\Rightarrow \bigcap\{a_i : i \in I\} \in \mathcal{A}$.

♠ EXERCISE. Prove (i)–(vi).

10.4. PROPOSITION. *If $X \subset \mathrm{P}(A)$ satisfies (i)–(iii) of 10.3 then there is a reflexive and symmetric $\sim$ on A such that $X = Coh((A, \sim))$.*
PROOF. Take $x \sim y := \{x, y\} \in X$. □

10.5. EXAMPLES

(a) For any set X, $X \equiv (X, =)$ is a ("discrete") web; the corresponding domain is called *flat*. We write $\underline{X}$ for the corresponding Girard domain. $Coh(X) = \{\emptyset\} \cup \{\{\alpha\} : \alpha \in X\}$, the empty set plus all singletons. For $\underline{\emptyset}$ we also write $\underline{0}$; $\emptyset$ is the only object in $\underline{0}$. For $\{0\}$ we also write $\underline{1}$; $Coh(\{0\}) = \{\emptyset, \{0\}\}$.

(b) For arbitrary sets X, Y consider $(X \times Y, \sim)$ with

$$(x, y) \sim (x', y') := \text{ if } x = x' \text{ then } y = y'.$$

$Coh(X \times Y, \sim)$ consists of the partial functions from X to Y.

(c) If $\mathbf{A} \equiv (A, \sim_A)$, $\mathbf{B} \equiv (B, \sim_B)$ are webs, then also

$$\mathbf{A} \sqcap \mathbf{B} = (A \mathbin{\dot{\cup}} B, \sim),$$

where
$$A \mathbin{\dot{\cup}} B := (\{0\} \times A) \cup (\{1\} \times B),$$
$$(0, \alpha) \sim (0, \alpha') \text{ iff } \alpha \sim_A \alpha',$$
$$(1, \beta) \sim (1, \beta') \text{ iff } \beta \sim_B \beta',$$
$$(0, \alpha) \sim (1, \beta) \text{ for all } \alpha \in A, \beta \in B.$$

Then

$$\mathrm{Coh}(\mathbf{A} \sqcap \mathbf{B}) \cong \mathrm{Coh}(\mathbf{A}) \times \mathrm{Coh}(\mathbf{B}),$$

that is to say there is a bijection between the left hand and the right hand side respecting $\subset$, since any $a \in \mathrm{Coh}(\mathbf{A} \sqcap \mathbf{B})$ can be uniquely split as

$$(\{0\} \times a_0) \cup (\{1\} \times a_1)$$

with $a_0 \in \mathrm{Coh}(\mathbf{A})$, $a_1 \in \mathrm{Coh}(\mathbf{B})$.

We shall next introduce some categories of Girard domains by defining suitable classes of morphisms.

10.6. DEFINITION. A map $F : \mathcal{A} \longrightarrow \mathcal{B}$ is said to be *monotone* if

(i) $a \subset a' \in \mathcal{A} \Rightarrow F(a) \subset F(a')$,

F is *continuous* if

(ii) if X is directed w.r.t. $\subset$ in $\mathcal{A}$, then $F(\bigcup X) = \bigcup\{F(b) : b \in X\}$. (The right hand side is defined for *any* F, but is not necessarily coherent.)

If F is continuous and moreover

(iii) $a \cup a' \in \mathcal{A} \Rightarrow F(a \cap a') = F(a) \cap F(a')$ (stability),

then F is *stable*, and if F also satisfies

(iv) If $X \subset \mathcal{A}$, and for all $b, c \in X \Rightarrow b \cup c \in \mathcal{A}$, then $F(\bigcup X) = \bigcup\{F(b) : b \in X\}$
(i.e., F commutes with arbitrary unions),

then F is *linear*. $\square$

COMMENTS.

(i) Continuity implies monotonicity: let $a \subset a'$, then $a \cup a' = a'$ and $\{a, a'\}$ is directed, so by continuity $F(a) = F(a \cup a') = F(a) \cup F(a')$.

(ii) Stability also implies monotonicity: if $a \subset a' \in \mathcal{A}$, then $F(a \cap a') = F(a) \cap F(a')$; stability does not imply continuity.

(iii) If we regard the posets $\mathcal{A}$, $\mathcal{B}$ as categories, then monotonicity makes F into a functor; continuity, stability and linearity express that F preserves directed colimits, pullbacks and arbitrary colimits respectively.

10.7. PROPOSITION. *Continuous, stable and linear maps have the following alternative characterizations.* $F : \mathcal{A} \longrightarrow \mathcal{B}$ *is continuous iff*

(ii)′ whenever $\beta \in F(a)$ there is a finite $a_0 \subset a$ such that $\beta \in F(a_0)$,

F is stable iff

(iii)′ whenever $\beta \in F(a)$ there is a least (necessarily finite) $a_0 \subset a$ such that $\beta \in F(a_0)$,

F is linear iff

(iv)′ whenever $\beta \in F(a)$, there is an $\alpha \in a$ such that $\beta \in F(\{\alpha\})$.

PROOF. To show the equivalence of continuity with condition (ii)′, it is sufficient to note that, since any coherent a is the directed limit of its finite subsets, $\beta \in F(a)$ iff $\beta \in a_0$ for some finite $a_0 \subset a$. We leave the proof of the remainder as an exercise. $\square$

♠ EXERCISE. Give the remainder of the proof.

10.8. DEFINITION. Dom is the category of Girard domains with as morphisms from $\mathcal{A}$ to $\mathcal{B}$ the continuous maps $F : \mathcal{A} \longrightarrow \mathcal{B}$.

Stab is the subcategory of Dom with stable morphisms only, and Lin the subcategory with linear morphisms only. We write $\mathrm{Dom}(\mathcal{A}, \mathcal{B})$, $\mathrm{Stab}(\mathcal{A}, \mathcal{B})$ etc. for the morphisms from $\mathcal{A}$ to $\mathcal{B}$. $\square$

10.9. PROPOSITION. *Dom, Stab, and Lin have finite products: terminal object is $\underline{0}$, the atomless Girard domain with trivial coherence relation. $\mathcal{X} \sqcap \mathcal{Y}$ is the categorical product of $\mathcal{X}$ and $\mathcal{Y}$, with projections and pairing $\langle , \rangle$*

$$\pi_0 a = \{\alpha : (0, \alpha) \in a\}, \quad \pi_1 a = \{\beta : (1, \beta) \in a\},$$

$$\langle F, G \rangle(a) = \{(0, \alpha) : \alpha \in F(a)\} \cup \{(1, \beta) : \beta \in G(a)\}. \quad \square$$

♠ EXERCISE. Prove this.

Our next aim is to show that in Stab and Lin the sets of morphisms $\mathrm{Stab}(\mathcal{A}, \mathcal{B})$ and $\mathrm{Lin}(\mathcal{A}, \mathcal{B})$ can be represented by objects $\mathcal{A} \to \mathcal{B}$, $\mathcal{A} \multimap \mathcal{B}$ respectively.

10.10. DEFINITION. Every continuous $F : \mathcal{A} \longrightarrow \mathcal{B}$ is determined by a subset $\mathrm{gr}(F)$ of $\mathcal{A}_{\mathrm{fin}} \times B$, the *graph* of F

$$\mathrm{gr}(F) := \{(a, \beta) : a \in \mathcal{A} \text{ and } \beta \in F(a)\},$$

and conversely, any set $X \subset \mathcal{A}_{\mathrm{fin}} \times B$ such that

(1)
$$\begin{cases} (a, \beta) \in X, a \subset a' \in \mathcal{A} \Rightarrow (a', \beta) \in X \\ (a, \beta), (a, \beta') \in X \Rightarrow \beta \sim \beta' \end{cases}$$

determines a continuous $F_X : \mathcal{A} \longrightarrow B$ by

$$F_X(a) := \{\beta : \exists a' \subset a (a', \beta) \in X\}.$$

If F is *stable*, F is determined by a subset $\mathrm{tr}(F)$ of $\mathrm{gr}(F)$, the *trace* of F:

$$\mathrm{tr}(F) = \{(a, \beta) : a \in \mathcal{A} \text{ a least set such that } \beta \in F(a)\}.$$

$\mathrm{tr}(F)$ is a set $X \subset \mathcal{A}_{\mathrm{fin}} \times B$ such that

(2)
$$\begin{cases} (a, \beta), (a', \beta') \in X, a \cup a' \in \mathcal{A} \Rightarrow \beta \sim \beta' \\ (a, \beta), (a', \beta) \in X, a \cup a' \in \mathcal{A} \Rightarrow a = a', \end{cases}$$

and conversely, any such set defines a stable morphism $F_{\mathrm{st}(X)}$ by

$$F_{\mathrm{st}(X)}(a) = \{\beta : \exists a' \subset a (a', \beta) \in X\}.$$

For a linear F all elements (a, β) of $\mathrm{tr}(F)$ are in fact of the form $(\{\alpha\}, \beta)$, so we can define the *linear trace* of F

$$\mathrm{ltr}_F := \{(\alpha, \beta) : \beta \in F(\{\alpha\})\}$$

which is a set $X \subset A \times B$ such that

(3)
$$\begin{cases} (\alpha, \beta), (\alpha', \beta') \in X, \alpha \sim \alpha' \Rightarrow \beta \sim \beta', \\ (\alpha, \beta), (\alpha', \beta) \in X, \alpha \sim \alpha' \Rightarrow \alpha = \alpha'. \end{cases}$$

Conversely, any set X satisfying these conditions determines a linear $F_{\mathrm{lin}(X)}$ by

$$F_{\mathrm{lin}(X)}(a) = \{\beta : \exists \alpha \in a((\alpha, \beta) \in X)\}. \qquad \square$$

10.11. DEFINITION. For webs $\mathbf{A}, \mathbf{B}$ let

$$\mathbf{A} \rightarrow \mathbf{B} := (\mathcal{A}_{\mathrm{fin}} \times B, \cong),$$

$$\mathbf{A} \multimap \mathbf{B} := (A \times B, \cong')$$

where

$(a, \beta) \cong (a', \beta') :=$
$\quad a \cup a' \in \mathcal{A} \Rightarrow (\beta \sim \beta' \text{ and } (\beta = \beta' \Rightarrow a = a')), \text{ and}$
$(\alpha, \beta) \cong' (\alpha', \beta') :=$
$\quad \alpha \sim \alpha' \Rightarrow (\beta \sim \beta' \text{ and } (\beta = \beta' \Rightarrow \alpha = \alpha')).$

The Girard domains $\mathcal{A} \rightarrow \mathcal{B}$, $\mathcal{A} \multimap \mathcal{B}$ corresponding to these webs are precisely the sets X satisfying (2) and (3) of 10.10 respectively. $\square$

10.12. PROPOSITION. **Stab** *is cartesian closed, with exponential* $\mathcal{A} \rightarrow \mathcal{B}$ *for every* $\mathcal{A}$ *and* $\mathcal{B}$; $\mathrm{ev} : (\mathcal{A} \rightarrow \mathcal{B}) \sqcap \mathcal{A} \longrightarrow \mathcal{B}$ *is defined by*

$$\mathrm{ev}(X \,\dot\cup\, a) = F_{\mathrm{st}(X)}(a),$$

and for any $F : \mathcal{A} \sqcap \mathcal{B} \longrightarrow \mathcal{C}$ *we define* $\mathrm{cur}(F) : \mathcal{A} \longrightarrow (\mathcal{B} \rightarrow \mathcal{C})$ *by*

$$\mathrm{cur}(F)(a) = \{(b, \gamma) : \gamma \in F(a \,\dot\cup\, b)\}.$$

PROOF. One has to verify that for all stable F, G

$$\mathrm{ev} \circ (\mathrm{cur}(F) \sqcap \mathrm{id}) = F,$$
$$\mathrm{cur}(\mathrm{ev} \circ (G \sqcap \mathrm{id})) = G,$$

where id and $\sqcap$ have the obvious definitions. $\square$

10.13. PROPOSITION. *The inclusion ordering on* $\mathcal{A} \rightarrow \mathcal{B}$ *corresponds to the stable ordering* $\lhd$ *on* $\mathsf{Stab}(\mathcal{A}, \mathcal{B})$, *where*

$$F \lhd G := \forall a, a' \in \mathcal{A}(a \subset a' \Rightarrow F(a) = F(a') \cap G(a)),$$

that is to say

$$F \lhd G \text{ iff } \mathrm{tr}(F) \subset \mathrm{tr}(G).$$

♠ EXERCISE. Prove the proposition.

10.14. PROPOSITION. **Lin** *is not cartesian closed.*

♠ EXERCISE. Prove the proposition.

10.15. PROPOSITION. **Dom** *is not cartesian closed.*
FIRST PROOF. We give a sketch only and leave the details to the reader.

An element a in a poset $(A, \leq)$ is said to be *compact* if for any directed $\{b_i : i \in I\}$ with $a \leq \bigvee_{i \in I} b_i$ there is an $i \in I$ with $a \leq b_i$. a is *left finite* if $\{b : b \leq a\}$ is finite. Then one successively shows:

(i) The compact elements of a Girard domain $\mathcal{A}$ are precisely the finite ones, and each compact element is left-finite.

(ii) There is a bijective correspondence between morphisms $\underline{0} \longrightarrow \mathcal{A}$ and $\mathcal{A}$.

(iii) There is a bijective correspondence between morphisms $\underline{1} \longrightarrow \mathcal{A}$ and ordered pairs (a, b) with $a \subset b$ in $\mathcal{A}$. ($\underline{1} = \{\emptyset, \{0\}\}$ and the only condition $F : \underline{1} \longrightarrow \mathcal{A}$ has to satisfy is $F(\emptyset) \subset F(\{0\})$; continuity is automatic).

(iv) There is a bijective correspondence between morphisms $\underline{1} \sqcap \mathcal{A} \longrightarrow \mathcal{B}$ and ordered pairs $F \leq G$ of morphisms from $\mathcal{A}$ to $\mathcal{B}$, where

$$F \leq G := \forall a \in \mathcal{A}(F(a) \subset G(a)).$$

(v) Assume $\mathcal{B}^{\mathcal{A}}$ to be the exponent of $\mathcal{A}$ and $\mathcal{B}$ in Dom, then to each $f : \underline{1} \sqcap \mathcal{A} \longrightarrow \mathcal{B}$ there is a unique $\mathrm{cur}(f)$ such that the diagram below commutes for suitable ev.

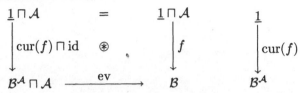

$\mathcal{B}^{\mathcal{A}}$ must now be isomorphic to $(\mathrm{Dom}(\mathcal{A}, \mathcal{B}), \leq)$.

(vi) Consider $F : \underline{\mathbb{N}} \longrightarrow \underline{1}$ with

$$F(\emptyset) = F(\{n\}) = \{1\}.$$

F is a compact element in $\mathrm{Dom}(\underline{\mathbb{N}}, \underline{1})$ under $\leq$, but F is not left-finite, hence $\mathrm{Dom}(\underline{\mathbb{N}}, \underline{1})$ cannot be isomorphic to a Girard domain. $\square$

SECOND PROOF. If we are willing to use a bit of general domain theory, there is another insightful proof. Let DCPO be the category of directed-complete partial orders (dcpo's) with continuous functions as morphisms. Then Lemma 1.21(iii) of Jung 1989 states the following. Let $\mathcal{C}$ be a cartesian closed full subcategory of DCPO. Then for any two objects A, B of $\mathcal{C}$ the exponential object $A \to B$ is isomorphic to $[A \to B]$, the set of continuous maps from A to B, pointwise ordered.

Observe that $[\underline{1} \to \underline{1}]$ is a linearly ordered dcpo with three elements. But the only possible Girard domain with three elements is isomorphic to $\{a, b\}, a \neq b$, with elements $\emptyset, \{a\}, \{b\}$, not linearly ordered under inclusion. Hence Dom cannot be cartesian closed. $\square$

10.16. EXAMPLE. We may define a non-stable map representing a kind of "parallel or"

$$F : \mathcal{B} \sqcap \mathcal{B} \longrightarrow \mathcal{B}$$

where $\mathcal{B}$ is the "space of booleans", a flat Girard domain with atoms t, f; F is given by

$$F(\{(0, f), (1, f)\}) = \{f\},$$
$$F(a) = t \text{ if } (0, t) \in a \text{ or } (1, t) \in a,$$
$$F(a) = \emptyset \text{ otherwise.}$$

For $a = \{(0, t), (1, t)\}$ there is no least $a_0 \subset a$ such that $t \in F(a_0)$, so F is not stable.

10.17. DEFINITION. Let $\mathcal{A}$, $\mathcal{B}$ be Girard domains with webs $\mathbf{A} \equiv (A, \sim)$, $\mathbf{B} := (A, \sim')$ respectively. Then the *tensor product* $\mathcal{A} \star \mathcal{B}$ of $\mathcal{A}$ and $\mathcal{B}$ is the domain generated by the web

$$\mathbf{A} \star \mathbf{B} := (A \times B, \approx), \text{ where } (x, y) \approx (x', y') := x \sim x' \text{ and } y \sim y'. \quad \square$$

10.18. PROPOSITION. *In* Lin $\multimap$ *is a closure w.r.t.* $\star$, *that is to say*

$$\text{Lin}(- \star \mathcal{A}, \mathcal{B}) \text{ is left adjoint to } \text{Lin}(-, \mathcal{A} \multimap \mathcal{B}),$$

or in a diagram: to each $f : \mathcal{C} \star \mathcal{A} \longrightarrow \mathcal{B}$ *there is a unique* $\text{cur}(f) \in \mathcal{C} \longrightarrow \mathcal{A} \multimap \mathcal{B}$ *such that for a fixed* $\text{ev}_{\mathcal{A}, \mathcal{B}}$ ($\text{ev}_{\mathcal{A}, \mathcal{B}}$ *natural in* $\mathcal{B}$)

$$
\begin{array}{ccccc}
\mathcal{C} \star \mathcal{A} & = & \mathcal{C} \star \mathcal{A} & & \mathcal{C} \\
\downarrow {\scriptstyle \text{cur}(f) \star \text{id}} & \circledast & \downarrow {\scriptstyle f} & & \downarrow {\scriptstyle \text{cur}(f)} \\
(\mathcal{A} \multimap \mathcal{B}) \star \mathcal{A} & \xrightarrow{\text{ev}_{\mathcal{A}, \mathcal{B}}} & \mathcal{B} & & \mathcal{A} \multimap \mathcal{B}
\end{array}
$$

commutes.

PROOF. For an arbitrary $f : \mathcal{C} \star \mathcal{A} \multimap \mathcal{B}$, the linear trace of F consists of triples $((\gamma, \alpha), \beta)$ and is related to the linear trace of $\text{cur}(f)$ by

$$((\gamma, \alpha), \beta) \in \text{ltr}_f \text{ iff } (\gamma, (\alpha, \beta)) \in \text{ltr}_{\text{cur}(f)}.$$

For $\text{ev}_{\mathcal{A}, \mathcal{B}}$ we can take the map G such that $\beta \in G(a)$ iff $((\alpha, \beta), \alpha) \in a$ for some $a \in (\mathcal{A} \multimap \mathcal{B}) \star \mathcal{A}$, so $\text{ltr}_G = \{(((\alpha, \beta), \alpha), \beta) : \alpha \in A, \beta \in B\}$. Now for $g \in \mathcal{B} \multimap \mathcal{C}, h \in \mathcal{C} \multimap \mathcal{D}$ we have

$$(\beta, \delta) \in \text{ltr}_{hg} \text{ iff } \exists \gamma \in C((\beta, \gamma) \in \text{ltr}_g \text{ and } (\gamma, \delta) \in \text{ltr}_h).$$

From this we readily see that

$$\text{ev} \circ ((\text{cur}(f) \star \text{id}) = f,$$

since the linear traces are the same. $\square$

10.19. The storage operator ! in Girard domains

DEFINITION. For any Girard domain $\mathcal{A}$ with web $(A, \sim)$, let $!\mathcal{A}$ be the domain with web $(\mathcal{A}_{\text{fin}}, \cong)$ with

$$a \cong b := a \cup b \in \mathcal{A}_{\text{fin}}. \quad \square$$

It is easy to see that we can write $\mathcal{A} \rightarrow \mathcal{B}$, the set of stable morphisms as represented by their traces under inclusion, as

$$!\mathcal{A} \multimap \mathcal{B}$$

where $\mathcal{X} \multimap \mathcal{Y}$ is the set of linear morphisms from $\mathcal{X}$ to $\mathcal{Y}$ as represented by their linear traces under inclusion. The discovery of this splitting of $\rightarrow$ into ! and $\multimap$ was a very important step towards the discovery of linear logic.

10.20. PROPOSITION. *Let* $\top$ *be interpreted as* $\underline{0}$, $\mathbf{1}$ *as* $\underline{1}$. *Then*

$$!\top \text{ is isomorphic to } \mathbf{1},$$

and for all $\mathcal{A}, \mathcal{B}$

$$!(\mathcal{A} \sqcap \mathcal{B}) \text{ is isomorphic to } !\mathcal{A} \star !\mathcal{B}.$$

10.21. PROPOSITION. *If we define for* $f : \mathcal{A} \longrightarrow \mathcal{B}$ *the corresponding* $!f : !\mathcal{A} \longrightarrow !\mathcal{B}$ *by*

$$!f(a) := f[a] \equiv \{f(x) : x \in a\},$$

where a *is a finite coherent set of* $!\mathcal{A}$, *i.e., a finite set* $\{a_1, \ldots, a_n\}$ *of* $a_i \in \mathcal{A}$, *with* $\bigcup_i a_i \in \mathcal{A}$, *then* ! *is a functor, and* $(!, \mathbf{r}, \mathbf{s})$ *is a comonad in the sense of category theory, where* $\mathbf{r} : ! \overset{\cdot}{\rightarrow} I$, $\mathbf{s} : ! \overset{\cdot}{\rightarrow} !!$ *and*

$$\mathbf{s}_A : !\mathcal{A} \longrightarrow !!\mathcal{A}$$

is the morphism mapping $a \in (!\mathcal{A})_{\text{fin}}$ *to* $\bigcup a \in \mathcal{A}_{\text{fin}}$, *and* $\mathbf{s}_A : !\mathcal{A} \longrightarrow !!\mathcal{A}$ *is simply defined as*

$$\mathbf{s}_A(\{a_1, \ldots, a_n\}) = \{\{a_1\}, \ldots, \{a_n\}\},$$

or on atoms: $\mathbf{s}_A(\{a_1\}) = \{\{a_1\}\}$ $\square$.

10.22. PROPOSITION. Lin *is a classical linear category with storage.*

We end with a table showing the interpretation of the various connectives in Girard domains, defined via correspondingly denoted operations on their webs.

TABLE 13
Interpretation of **CLL** in Girard domains

Let $\mathbf{X} \equiv (X, \sim)$ and $\mathbf{Y} \equiv (Y, \sim')$ be webs. We define
Negation

$$\sim\mathbf{X} := (X, \approx) \text{ with } x \approx y := x \not\sim y \text{ or } x = y$$

Tensor

$$\mathbf{X} \star \mathbf{Y} := (X \times Y, \approx) \text{ with } (x, y) \approx (x', y') := x \sim x' \text{ and } y \sim' y'$$

Par

$$\mathbf{X} + \mathbf{Y} \equiv (X \times Y, \approx) \text{ with}$$

$$(x, y) \approx (x', y') := (x, y) = (x', y') \text{ or}$$
$$(x \sim x' \text{ and } x \neq x') \text{ or } (y \sim' y' \text{ and } y \neq y')$$

Linear implication

$$\mathbf{X} \multimap \mathbf{Y} := (X \times Y, \approx) \text{ with}$$

$$(x, y) \approx (x', y') := x \sim x' \Rightarrow (y \sim' y' \text{ and } (y = y' \Rightarrow x = x'))$$

Conjunction

$$\mathbf{X} \sqcap \mathbf{Y} := (X \dot\cup Y, \approx), \text{ where}$$

$$X \dot\cup Y := (\{0\} \times X) \cup (\{1\} \times Y) \text{ and}$$

$$(a, x) \approx (b, y) := ((a = b = 0 \text{ and } x \sim y) \text{ or}$$
$$(a = b = 1 \text{ and } x \sim' y) \text{ or}$$
$$(a = 0 \text{ and } b = 1) \text{ or}$$
$$(a = 1 \text{ and } b = 0))$$

Disjunction

$$\mathbf{X} \sqcup \mathbf{Y} := (X \dot\cup Y, \approx) \text{ with}$$

$$(a, x) \approx (b, y) := (a = b = 0 \text{ and } x \sim y) \text{ or } (a = b = 1 \text{ and } x \sim' y)$$

Constants. $\top$ and $\bot$ are both interpreted by $\underline{0}$, the Girard domain without atoms. $\mathbf{0}$ and $\mathbf{1}$ are both interpreted by $\underline{1}$, the Girard domain with a single atom.
Exponentials

$$!\mathbf{X} := (\text{Fincoh}(\mathbf{X}), \approx) \text{ with } a \approx b := a \cup b \in \text{Coh}(\mathbf{X})$$

$$?\mathbf{X} := (\text{Fincoh}(\sim\mathbf{X}), \approx) \text{ with } a \approx b := a = b \text{ or } a \cup b \notin \text{Coh}(\sim\mathbf{X})$$

11 Coherence in symmetric monoidal categories

11.1. The present chapter, devoted to the coherence problem for symmetric monoidal categories (corresponding to the tensor fragment of linear logic), is not absolutely necessary for what follows. In Chapter 9 we saw how to construct several types of free categories over a discrete graph (i.e., a graph without edges). All the arrows in such a category are standard arrows: they exist in any category of the same type. The coherence problem for the given type of category is simply this: which standard arrows (more accurately: which terms denoting standard arrows) are the same in the free category, and hence everywhere (more accurately: denote the same arrow in the free category), and how many arrows are there between two given objects in the free category?

Following MacLane (1963, 1971), we shall present the solution for symmetric monoidal categories. The answer is very simple in this case, since we have the following theorem:

11.2. THEOREM. *Each object of the free SMC-category is a tensor product of generators and* **1**. *Let* $x_1, x_2, \ldots x_p$, $y_1, y_2, \ldots, y_q$ *be the generators appearing in objects* A *and* B *respectively, in that order. There are morphisms from* A *to* B *iff* $x_1, x_2, \ldots x_p$ *and* $y_1, y_2, \ldots, y_q$ *are the same as multisets; there are as many morphisms from* A *to* B *as there are permutations of 1,2,...,p leaving* $x_1, x_2, \ldots, x_p$ *invariant.*

102

In a free monoidal category there is a unique morphism from A to B iff $x_1, x_2, \ldots x_p$ and $y_1, y_2, \ldots, y_q$ are the same as sequences.

This basic case of the coherence problem is of interest to us as it shows that the transition from sequents $\Gamma \Rightarrow A$ to 1-sequents $B \Rightarrow A$ is rather harmless, not only logically, but also from a type-theoretic point of view. The proof requires a number of lemmata and will start with a result for premonoidal categories (without λ, γ and the unit object); this is then extended to symmetric premonoidal categories (addition of γ) and finally to to SMC's (addition of λ and $\mathbf{1}$).

11.3. DEFINITION. A *word* of *length* n over p generators $x_1, \ldots, x_p$ is defined inductively by

(i) $\mathbf{1}$ is a word of length 0; x_i is a word of length 1.

(ii) If v,w are words of length n and m respectively, then $(v \star w)$ is a word of length $n + m$.

A word not containing $\mathbf{1}$ is called *pure*. The *rank* $\rho(w)$ of a word w is defined by

(iii) $\rho(\mathbf{1}) = 0$, $\rho(x_i) = 0$,

(iv) $\rho(v \star w) = \rho(v) + \rho(w) + \text{length}(w) - 1$.

The rank measures the number of pairs of parentheses not starting at the left, so $\rho(v) = 0$ means that all pairs of parentheses start at the left. □

11.4. DEFINITION. Let $\mathcal{C}_p$ be the free premonoidal category in p generators $x_1, \ldots, x_p$, and let G_p be the graph with as nodes all objects of length p in which $x_1, \ldots, x_p$ occur in that order from left to right, and as arrows (= directed edges) *basic* arrows constructed from instances of α, α^{-1}, and identity arrows, that is to say

(i) any instance of α, α^{-1} is basic,

(ii) if β is basic, then so are $\beta \star \text{id}$, $\text{id} \star \beta$.

A basic arrow is *directed* if it contains an instance of α, and *antidirected* if it contains an instance of α^{-1}.
The nodes of G_p are part of the objects of $\mathcal{C}_p$. □

11.5. LEMMA. *All paths from v to w in G_p represent the same arrow in $\mathcal{C}_p$.*

PROOF. Let $w^{(p)}$ be the unique object in G_p with association to the left, i.e.,

$$w^{(p)} = (\ldots((x_1 \star x_2) \star x_3)\cdots \star x_p)$$

For an arbitrary w in G_p there is a canonical directed path from w to $w^{(p)}$ obtained by systematically applying α, starting on the outside.

In order to prove that all paths from v to w in G_p represent a unique arrow in C_p, it suffices to show that any two directed paths from v to $w^{(p)}$ represent the same arrow. This is seen by considering, for any path in G_p

$$v \equiv v_0 \xrightarrow{\ f_1\ } v_1 \xrightarrow{\ f_2\ } v_2 \xrightarrow{\ f_3\ } \cdots \xrightarrow{\ f_n\ } v_n \equiv w$$

the subgraph obtained by connecting each v_i to $w^{(p)}$ by the canonical path; we leave this to the reader.

The fact that any two directed paths from v to $w^{(p)}$ represent the same arrow is proved by induction on the rank of v.

Let two different directed paths starting from v, with rank n, start with the basic arrows $\beta_0 : v \longrightarrow v_0$, $\beta_1 : v \longrightarrow v_1$; both β_0 and β_1 decrease the rank. So if we can construct z with directed arrows $v_0 \longrightarrow z$, $v_1 \longrightarrow z$ making the diamond from v to z commutative, we are done.

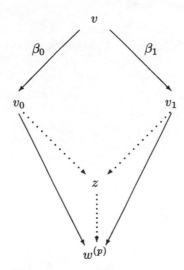

Let $v \equiv (u \star w)$. We distinguish cases.

   Case 1. If $\beta_0 \equiv \beta_1$ the solution is trivial. If $\beta_0 \not\equiv \beta_1$, there are three possibilities for β_0.

(a) $\beta_0 = \beta' \star \mathrm{id}_w$ (β_0 "acts" in the first factor),

(b) $\beta_0 = \mathrm{id}_u \star \beta''$ (β_0 "acts" in the second factor),

(c) $\beta_0 = \alpha_{u,s,t}$ where $v \equiv (u \star w) \equiv (u \star (s \star t))$.

The same three possibilities exist for β_1. This gives rise to the following cases.

Case 2. Both β_0 and β_1 act inside the same factor: use induction on the *length*.

Case 3. β_0 acts inside u, β_1 acts inside w. then the result follows since the diagram

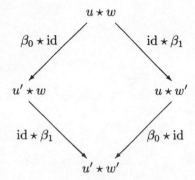

commutes by the functorial character of $\star$.

Case 4. Either β_0 or β_1 (say β_0) is an instance of α; let $\beta_0 = \alpha_{u,s,t}$ as under (c) above. $\beta_0 \neq \beta_1$, so β_1 must act inside u or inside w. This yields three subcases.

Subcase 4.1. β_1 acts inside u. The result follows since the diamond

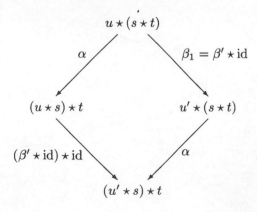

commutes because of the naturality of $\star$.

Subcase 4.2. β_1 acts inside $w = s \star t$, in fact inside s or inside t; we may construct a diamond as before.

Subcase 4.3. β_1 acts inside w, but not inside s or t; then β_1 is itself an instance of α, $t \equiv t' \star t''$, and the diamond to be constructed starts with

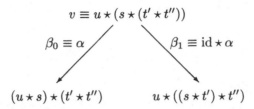

$$v \equiv u \star (s \star (t' \star t''))$$

$$\beta_0 \equiv \alpha \qquad\qquad \beta_1 \equiv \mathrm{id} \star \alpha$$

$$(u \star s) \star (t' \star t'') \qquad\qquad u \star ((s \star t') \star t'')$$

The diagram may now be completed into a commuting diamond by the pentagon condition.

11.6. COROLLARY. *The lemma establishes Theorem 11.2 for the free premonoidal categories.*

Now we shall extend the argument.

11.7. DEFINITION. Let $\mathcal{C}_p^*$ be the free symmetric premonoidal category in p generators $x_1, \ldots, x_p$, and let G_p^* be the graph with as vertices all tensor products constructed from $x_1, \ldots, x_p$ in which each x_i occurs exactly once, and with edges the basic arrows generated by

(i) instances of α, α^{-1}, γ, γ^{-1} are basic arrows;

(ii) if β is a basic arrow, then so are $\mathrm{id} \star \beta$, $\beta \star \mathrm{id}$. $\square$

The vertices of G_p^* may be distributed over $p!$ boxes, one box for each permutation of $x_1, \ldots, x_p$ (i.e., in all objects in box ξ the x_i appear in the order $x_{\xi 1}, \ldots, x_{\xi p}$). Inside each box is a subgraph isomorphic to G_p, and we know by the preceding result that each closed (= circular) path in a box is the identity arrow.

An edge between boxes is an expansion of an arrow $\gamma_{A,B}$, for example

$$\mathrm{id}_{x_2} \star \gamma_{x_3 \star x_5, x_4 \star x_1} : x_2 \star ((x_3 \star x_5) \star (x_4 \star x_1)) \longrightarrow x_2 \star ((x_4 \star x_1) \star (x_3 \star x_5)).$$

Now consider the hexagon

$$A \star (B \star C) \xrightarrow{\;\alpha\;} (A \star B) \star C \xrightarrow{\;\gamma\;} C \star (A \star B)$$

$$\Big\downarrow \mathrm{id} \star \gamma \qquad\qquad\qquad\qquad\qquad \Big\downarrow \alpha$$

$$A \star (C \star B) \xrightarrow{\;\alpha\;} (A \star C) \star B \xrightarrow{\;\gamma \star \mathrm{id}\;} (C \star A) \star B$$

Interchanging the block $(A \star B)$ with C can be replaced by two interchanges of single letters B with C, and A with C. As a result, repeated use of the hexagon permits us to replace any interchange of generators by interchange of adjacent positions, i.e., permutations $\sigma_i = (i, i+1)$ $(1 \leq i \leq p)$ in the usual cycle notation. That is to say, each path β in G_p^* can be transformed into another path β' representing the *same* arrow in the category, such that the only instances of γ in β' are σ_i's.

Now the symmetric group $S(p)$ is generated by the σ_i, and hence there is at least one path between any two vertices in G_p^*.

Any *closed* path thereby gives rise to an identity of the form

$$\sigma_{n(0)} \sigma_{n(1)} \cdots \sigma_{n(k)} = 1$$

By a theorem from group theory (Burnside 1912, Note C) all such identities follow from the following set

$$\sigma_i^2 = 1 \quad (1 \leq i \leq p)$$
$$\sigma_i \sigma_j \sigma_i \sigma_j = 1 \quad (1 \leq i < j - 1 < p - 1)$$
$$(\sigma_i \sigma_{i+1})^3 = 1 \quad (1 \leq i < p - 1)$$

However, each relation corresponds to a closed path which is the identity.

(A) $\sigma_i^2 = 1$ corresponds to $\gamma \circ \gamma = \mathrm{id}$.

(B) For the second relation, note the commutativity of the following diagram (dropping $\star$ in the notation for objects):

$$(AB)(CD) \xrightarrow{\;\mathrm{id} \star \gamma\;} (AB)(DC)$$

$$\Big\downarrow \gamma \star \mathrm{id} \qquad \circledast \qquad \Big\downarrow \gamma \star \mathrm{id}$$

$$(BA)(CD) \xrightarrow{\;\mathrm{id} \star \gamma\;} (BA)(DC)$$

(C) For the third relation consider the following polygon with two dotted arrows inserted.

$$A(BC) \xrightarrow{\text{id} \star \gamma} A(CB) \xrightarrow{\alpha} (AC)B \xrightarrow{\gamma \star \text{id}} (CA)B$$

$$A(BC) \xrightarrow{\alpha} (AB)C \xrightarrow{\gamma \star \text{id}} (BA)C \xrightarrow{\alpha} B(AC)$$

$$(CA)B \xrightarrow{\alpha} C(AB) \xrightarrow{\text{id} \star \gamma} C(BA) \xrightarrow{\alpha} (CB)A$$

$$(AB)C \xcdots{\gamma} C(AB)$$

$$(BA)C \xcdots{\gamma} C(BA)$$

$$B(AC) \xrightarrow{\text{id} \star \gamma} B(CA) \xrightarrow{\alpha} (BC)A \xrightarrow{\gamma \star \text{id}} (CB)A$$

Then the middle rectangle commutes because γ is a natural transformation; the upper and lower rectangle are instances of the hexagon condition. As a result, the outer rectangle commutes.

An application of a basic identity corresponds to a substitution of one path by another which represents the same arrow. Note that any diagram obtained from one of the basic diagrams replacing a basic arrow β not involving γ by a path not involving γ, then the path is an arrow equal to β. As a result we have

11.8. PROPOSITION. *Between any two vertices of G_p^* there exists a unique arrow in $\mathcal{C}^*$.*

We now extend our result to symmetric monoidal categories. For this we need a lemma.

11.9. LEMMA. *From the hexagon condition and the triangle condition we obtain commutativity of the following diagram*

$$a \star (\mathbf{1} \star c) \xrightarrow{\alpha} (a \star \mathbf{1}) \star c \xrightarrow{\gamma \star id} (\mathbf{1} \star a) \star c$$

$$\downarrow id \star \lambda \qquad \circledast \qquad \downarrow \lambda \star id$$

$$a \star c \qquad = \qquad a \star c$$

PROOF. Consider the hexagon condition for $\mathbf{1} \star (c \star a)$. Fill the hexagon with

(1) commuting triangle $(\lambda \star \text{id}) \circ \alpha = \lambda : \mathbf{1}(ca) \longrightarrow ca$,

(2) commuting square $\gamma \circ (\text{id} \star \lambda) = (\lambda \star \text{id}) \circ \gamma : a(\mathbf{1}c) \to ca$,

(3) commuting square $\lambda \circ (\text{id} \star \gamma) = \gamma \circ \lambda : \mathbf{1}(ca) \to ac$,

(4) commuting triangle $(\lambda \star \text{id}) \circ \alpha = \lambda : \mathbf{1}(ac) \to ac$,

based on the triangle condition and the naturality of γ, λ. We obtain a commuting square

$$(\lambda \star \text{id}) \circ (\gamma \circ \text{id}) \circ \alpha = ((\lambda \circ \gamma) \star \text{id}) \circ \alpha = \text{id} \star \lambda. \qquad \square$$

11.10. DEFINITION. Let $\mathcal{C}^{**}$ be the free symmetric monoidal category, let $x_1, \ldots, x_p$ be p generators, and let G_p^{**} be defined as before except that the vertices may now contain, besides exactly one occurrence of each generator, also arbitrarily many occurrences of $\mathbf{1}$.

Edges are basic arrows; now arrows are basic if they are constructed by

(i) α, α^{-1}, γ, γ^{-1}, λ, λ^{-1}, ρ, ρ^{-1} (where $\rho_A = \lambda_A \circ \gamma_{A,1}$) are basic arrows.

(ii) If ϕ is a basic arrow, then so is $\text{id} \star \phi$, $\phi \star \text{id}$. $\square$

In order to prove that for any two paths from v to w in G_p^{**} correspond to the same arrow, we note that without loss of generality we may assume that $w \in G_p^*$, i.e., does not contain occurrences of $\mathbf{1}$ (simply apply to an arbitrary w basic arrows with λ or ρ until all $\mathbf{1}$'s are removed). Then we are done if we can show:

(i) any path is equivalent to a path where first all factors $\mathbf{1}$ are eliminated, and

(ii) any two different factors $\mathbf{1}$ can be removed in any order.

(ii) follows by the naturality of λ (exercise). As to (i), we have to show that α or γ followed by a λ- or ρ-expansion can be replaced by a λ- or ρ-expansion followed by α's and γ's. Most cases follow by naturality.

(a) $\lambda \circ \gamma$ is replaced by ρ, by definition. $\rho \circ \gamma$ is identical with λ.

(b) A sequence $\mathbf{1}\star(B\star C) \xrightarrow{\ \alpha\ } (\mathbf{1}\star B)\star C \xrightarrow{\ \lambda \star \text{id}\ } B\star C$ is replaced by λ (the triangle condition).

(c) A sequence $A\star(\mathbf{1}\star C) \xrightarrow{\ \alpha\ } (A\star\mathbf{1})\star C \xrightarrow{\ \rho \star \text{id}\ } A\star C$ is replaced by $\text{id} \star \lambda$.

(d) For $A \star (B \star 1) \xrightarrow{\alpha} (A \star B) \star 1 \xrightarrow{\rho} A \star B$ consider the following diagram:

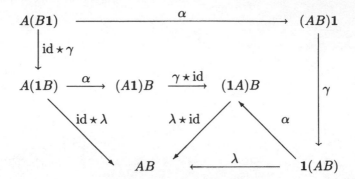

The outer polygon is commuting; inserting the inner arrows, we have on top an instance of the hexagon condition; the right triangle is the triangle condition, and the left triangle is the derived condition of the preceding lemma.

11.11.   Coherence problems for closed categories have been investigated in a.o. Kelly and MacLane 1971, Mints 1976 and Jay 1990. The situation now changes drastically: there are not always arrows between words based on the same multiset of generators in a free closed category . The existence of some arrow between two given words is decidable.

12 The storage operator as a cofree comonoid

12.1. This chapter is a digression from the main line of our exposition, and is not needed in later chapters. In Chapter 9 we encountered a "minimal" interpretation of ! in categorical semantics, namely as a comonad with two additional properties. In his thesis (1989), Lafont proposed another interpretation (also, independently, proposed in da Paiva 1989): $!A$ is the cofree comonoid over A ("cogenerated by A").

In the phase structures of Chapter 8 the operation ! corresponds under this interpretation to

$$!x := \prod_{n \in \mathbb{N}} (\mathbf{1} \sqcap x)^n$$

where $y^0 \equiv \mathbf{1}$, $y^{n+1} \equiv y^n \star y$.

In the category Lin, Lafont's interpretation of $!A$ becomes the least fixed point of the equation $X \cong \mathbf{1} \sqcap A \sqcap (X \star X)$. Below we first present the abstract categorical interpretation according to Lafont, and then turn to concrete examples. In our exposition we used Asperti 1990.

12.2. DEFINITION. Let $\mathcal{C}^{-} \equiv (\mathcal{C}, \star, \alpha, \gamma, \lambda, \mathbf{1})$ be an SMC. A triple (A, ϵ, δ) with $A \in \mathcal{C}$, $\epsilon : A \longrightarrow \mathbf{1}$, $\delta : A \longrightarrow A \star A$ is a (commutative) *comonoid* if the following diagrams commute:

$$
\begin{array}{ccccc}
AA & \xleftarrow{\;\delta\;} & A & \xrightarrow{\;\delta\;} & AA \\
\downarrow{\scriptstyle \mathrm{id}\star\delta} & & \circledast & & \downarrow{\scriptstyle \delta\star\mathrm{id}} \\
A(AA) & & \xrightarrow{\;\;\alpha\;\;} & & (AA)A
\end{array}
$$

$$
\begin{array}{ccc}
A & = & A \\
\downarrow{\scriptstyle \delta} & \circledast & \downarrow{\scriptstyle \delta} \\
AA & \xrightarrow{\;\gamma\;} & AA
\end{array}
\qquad
\begin{array}{ccc}
AA & \xrightarrow{\;\epsilon\star\mathrm{id}\;} & 1A \\
\uparrow{\scriptstyle \delta} & \circledast & \downarrow{\scriptstyle \lambda} \\
A & = & A
\end{array}
$$

ϵ is called the *eraser* and δ the *duplicator* of the comonoid. $\square$

12.3. REMARK. From the definition it is obvious that also

$$
\rho \circ (\mathrm{id} \star \epsilon) \circ \delta = \mathrm{id}_A,
$$

where ρ as before is $\lambda \circ \gamma$.

12.4. DEFINITION. Let $\mathcal{C}$ be an SMC as before. Comon$(\mathcal{C})$ is a category with as objects the comonoids in $\mathcal{C}$. An arrow

$$
f : (A, \epsilon, \delta) \longrightarrow (A', \epsilon', \delta')
$$

in Comon$(\mathcal{C})$ is an arrow of $\mathcal{C}$ such that

$$
\delta' \circ f = (f \star f) \circ \delta, \quad \epsilon' \circ f = \epsilon,
$$

i.e., the following diagrams commute:

$$
\begin{array}{ccc}
A & \xrightarrow{\;\delta\;} & AA \\
\downarrow{\scriptstyle f} & \circledast & \downarrow{\scriptstyle f\star f} \\
A' & \xrightarrow{\;\delta'\;} & A'A'
\end{array}
\qquad
\begin{array}{ccc}
A & \xrightarrow{\;\epsilon\;} & 1 \\
\downarrow{\scriptstyle f} & \circledast & \downarrow{\scriptstyle \mathrm{id}_1} \\
A' & \xrightarrow{\;\epsilon'\;} & 1
\end{array}
$$

$\square$

The following lemma will be needed later.

12.5. LEMMA

(i) $(1, id_1, \lambda^{-1})$ is a comonoid;

(ii) if (A, ϵ, δ) and (A', ϵ', δ') are comonoids, then so is

$$(A \star A', \lambda \circ (\epsilon \star \epsilon'), m \circ (\delta \star \delta')),$$

where m_{BCDE} is an isomorphism constructed from $\alpha, \alpha^{-1}, \gamma,$ id such that

$$m : (B \star C) \star (D \star E) \longrightarrow (B \star D) \star (C \star E).$$

(Specifically, we can take $m = \alpha^{-1} \circ (\alpha \star id) \circ ((id \star \gamma) \star id) \circ (\alpha^{-1} \star id) \circ \alpha).$
♠ EXERCISE. Prove the lemma.

12.6. DEFINITION. Let $\mathcal{C}$ be an SMC, $B_0 \in \mathcal{C}$. $(B, \mathbf{r}, \mathbf{e}, \mathbf{d})$ is a *cofree comonoid over B* if

(i) $(B, \mathbf{e}, \mathbf{d}) \in \text{Comon}(\mathcal{C})$, $\mathbf{r} : B \longrightarrow B_0$;

(ii) For each monoid (A, ϵ, δ) and $\phi : A \longrightarrow B_0$ there is a unique morphism $g \equiv g(\phi, \epsilon, \delta) : A \longrightarrow B$ such that

$$\mathbf{r} \circ g = \phi, \quad \mathbf{e} \circ g = \epsilon, \quad \mathbf{d} \circ g = (g \star g) \circ \delta.$$

The arrow $\mathbf{r}$ is called the *read*-operator of the cofree comonoid. □

REMARK. The definition may alternatively be expressed by saying that $(\mathbf{r}, (B, \mathbf{e}, \mathbf{d}))$ is a *universal arrow from F to B_0*, where $F : \text{Comon}(\mathcal{C}) \longrightarrow \mathcal{C}$ is the forgetful functor; i.e., for each $f : F(A) \longrightarrow B_0$ there is a unique arrow $\phi : (A, \epsilon, \delta) \longrightarrow (B, \mathbf{e}, \mathbf{d})$ such that

$$\mathbf{r} \circ F\phi = f.$$

12.7. DEFINITION. $\mathcal{C} \equiv (\mathcal{C}, \star, \mathbf{1}, \alpha, \gamma, \lambda, !, \mathbf{r}, \mathbf{e}, \mathbf{d})$ is an SMC *with free storage* (SMCFS) if $(\mathcal{C}, \star, \mathbf{1}, \alpha, \gamma, \lambda)$ is an SMC and $(!A, \mathbf{r}_A, \mathbf{e}_A, \mathbf{d}_A)$ is a cofree comonoid over A for each $A \in \mathcal{C}$. We write ILCFS (CLCFS) for an ILC (CLC) with free storage. □

REMARK. By a familiar argument from basic category theory this is equivalent to the statement that the forgetful functor $F : \text{Comon}(\mathcal{C}) \longrightarrow \mathcal{C}$ has a right adjoint ! which on objects A is given by definition as $!A$, and on arrows $f : A \longrightarrow B$ by

$$!f := g(f \circ \mathbf{r}_A, \mathbf{e}_A, \mathbf{d}_A) : !A \longrightarrow !B.$$

The verification is easy with (i) of the following lemma.

12.8. LEMMA. *Let C be an SMC. Then*

(i) Let $f : (A', \epsilon', \delta') \longrightarrow (A, \epsilon, \delta)$ be an arrow in $Comon(C)$, then

$$g(\phi \circ f, \epsilon', \delta') = g(\phi, \epsilon, \delta) \circ f.$$

(ii) With $f : A \longrightarrow B$, $h : C \longrightarrow A$, $(C, \epsilon, \delta) \in Comon(C)$

$$g(f \circ h, \epsilon, \delta) = {!}f \circ g(h, \epsilon, \delta).$$

(iii) If $(C, !)$ is an SMCFS, then

$$\mathbf{r} : {!} \overset{\cdot}{\to} I_C, \quad \mathbf{e} : {!} \overset{\cdot}{\to} \mathbf{1}, \quad \mathbf{d} : {!} \overset{\cdot}{\to} {!} \star {!}$$

are natural transformations, where $\mathbf{r} \equiv \langle \mathbf{r}_A \rangle_A$, $\mathbf{e} \equiv \langle \mathbf{e}_A \rangle_A$, $\mathbf{d} \equiv \langle \mathbf{d}_A \rangle_A$, I_C is the identity functor on C, and $\mathbf{1}$ stands for the constant functor from C to $\mathbf{1}$.

PROOF. As to (i), note that for $g \equiv g(\phi, \epsilon, \delta)$:

$$\mathbf{r}_B \circ g \circ f = \phi \circ f,$$
$$\mathbf{e}_B \circ g \circ f = \epsilon \circ f = \epsilon',$$
$$\mathbf{d}_B \circ g \circ f = (g \star g) \circ \delta \circ f = (g \star g) \circ (f \star f) \circ \delta' =$$
$$((g \circ f) \star (g \circ f)) \circ \delta'.$$

and because of the uniqueness condition on g it follows that $g(\phi \circ f, \epsilon', \delta) = g(\phi, \epsilon, \delta) \circ f$.

(ii), (iii) are left to the reader. $\square$

♠ EXERCISE. Prove (ii), (iii) of the lemma.

CONVENTION. Since in the context the second and third argument in g are automatically determined, we simply write $g(\phi)$ for $g(\phi, \epsilon, \delta)$ most of the time. $\square$

REMARK. The adjunction $(F, !, g^{-1})$ where

$$g_{A,B} : C(F(A, \epsilon, \delta), B) \longrightarrow Comon(C)((A, \epsilon, \delta), ({!}B, \mathbf{e}_B, \mathbf{r}_B))$$

is natural in A and B, may also be written as $(F, !, \eta, \eta')$ where η and η' are unit and counit of the adjunction, that is to say $\eta : I_{Comon(C)} \overset{\cdot}{\to} {!}F$ and $\eta' : F! \overset{\cdot}{\to} I_C$ satisfy

$$\eta_A = g_{A,!A}^{-1}(\mathrm{id}_{!A}) : {!}A \longrightarrow {!}A,$$

$$\eta'_A = g_{!A,A}(\mathrm{id}_{(!A, \mathbf{e}_A, \mathbf{d}_A)}) : {!}A \longrightarrow A$$

from which we see that $\eta_A = \mathrm{id}_{!A}$, $\eta'_A = \mathbf{r}_A$.

12.9. LEMMA. $(!, \mathbf{s}, \mathbf{r})$ *is a comonad, where* $\mathbf{s} : ! \rightarrow !!$ *is the natural transformation with* $\mathbf{s}_A = !\eta_A$.

PROOF. Quite generally, if (F, G, η, eta') is an adjunction, $(GF, \eta, G\eta'F)$ is a monad and $(FG, G\eta F, \eta')$ is a comonad (see e.g., MacLane 1971). This may also be checked by direct verification for the particular example considered here. □

♠ EXERCISE. Give a direct verification that $(!, \mathbf{s}, \mathbf{r})$ as defined above is a comonad.

12.10. THEOREM. *An ILCFS* $\mathcal{C}$ *is an ILC with storage.*

PROOF. We have to construct natural isomorphisms $\mathbf{p}$ and $\mathbf{t}$ such that

$$\mathbf{p}_{A,B} : !(A \sqcap B) \longrightarrow !A \star !B,$$
$$\mathbf{t} : !\top \longrightarrow \mathbf{1}.$$

We consider the more difficult case of $\mathbf{p}$ and leave the other case as an exercise. We put

$$\mathbf{p}_{A,B} := (!\pi_0 \star !\pi_1) \circ \mathbf{d}_{A \sqcap B}.$$

It has to be checked that $\mathbf{p}$ is a comonoid morphism, i.e.,

$$\delta \circ \mathbf{p}_{A,B} = (\mathbf{p}_{A,B} \star \mathbf{p}_{A,B}) \circ \mathbf{d}_{A \sqcap B},$$
$$\epsilon \circ \mathbf{p}_{A,B} = \mathbf{e}_{A,B},$$

where $(!A \star !B, \epsilon, \delta)$ is the comonoid of Lemma 12.8, i.e.,

$$\delta \equiv m \circ (\mathbf{d}_A \star \mathbf{d}_B), \quad \epsilon \equiv \lambda \circ (\mathbf{e}_A \star \mathbf{e}_B).$$

We check the first of these. First observe that the following diagram commutes:

$$
\begin{array}{ccccc}
!A!A & \xrightarrow{\mathbf{d}_A \star \mathbf{d}_A} & (!A!A)(!A!A) & \xrightarrow{\quad m \quad} & (!A!A)(!A!A) \\
{\scriptstyle !f \star !g}\downarrow & & {\scriptstyle (!f \star !f) \star (!g \star !g)}\downarrow & & {\scriptstyle (!f \star !g) \star (!f \star !g)}\downarrow \\
!B!C & \xrightarrow{\mathbf{d}_B \star \mathbf{d}_C} & (!B!B)(!C!C) & \xrightarrow{\quad m \quad} & (!B!C)(!B!C)
\end{array}
$$

(the right rectangle commutes since m is natural). Also we have commutation of

$$
\begin{array}{ccccc}
!A!A & \xleftarrow{\quad \mathbf{d}_A \quad} & !A & \xrightarrow{\quad \mathbf{d}_A \quad} & !!A \\
{\scriptstyle \mathbf{d}_A \star \mathbf{d}_A}\downarrow & & & & {\scriptstyle \mathbf{d}_A \star \mathbf{d}_A}\downarrow \\
(!A!A)(!A!A) & & \xrightarrow{\quad\quad m \quad\quad} & & (!A!A)(!A!A)
\end{array}
$$

This follows from the comonoid properties of $!A$ (recall that m is constructed from id, α, α^{-1}, γ). Instead of an ad hoc verification, one can also see this as a special instance of the following property. Let B be any tensor product of factors 1 and A, A part of a comonoid (A, ϵ, δ). For each B we define μ_B, by induction on the construction of B: $\mu_1 := \epsilon$, $\mu_A := \mathrm{id}_A$, $\mu_{B_1} \star \mu_{B_2} = (\mu_{B_1} \star \mu_{B_2}) \circ \delta$. Now if $\xi : B \longrightarrow B'$ is any canonical arrow from B to B' in the sense of Chapter 11 (i.e., built from id, λ, λ^{-1}, γ, α, α^{-1}) then $\mu_{B'} = \xi \circ \mu_B$ (cf. the dual lemma in MacLane 1971). We may then combine the preceding commuting diagrams and obtain commutativity of

$$
\begin{array}{ccccc}
C & \xleftarrow{\quad \mathbf{d} \quad} & !(A \sqcap B) & \xrightarrow{\quad \mathbf{d} \quad} & C \\[2pt]
{\scriptstyle \mathbf{d} \star \mathbf{d}} \downarrow & & \circledast & & \downarrow {\scriptstyle \mathbf{d} \star \mathbf{d}} \\[2pt]
CC & \xrightarrow{\hspace{4cm} m \hspace{4cm}} & & & CC \\[2pt]
{\scriptstyle !\pi_0 \star !\pi_1} \downarrow & & \circledast & & \downarrow {\scriptstyle (!\pi_0 \star !\pi_1) \star (!\pi_0 \star !\pi_1)} \\[2pt]
!A!B & \xrightarrow{\ \mathbf{d} \star \mathbf{d}\ } (!A!A)(!B!B) & \xrightarrow{\quad m \quad} & (!A!B)(!A!B)
\end{array}
$$

where $\mathbf{d} \equiv \mathbf{d}_{A \sqcap B}$, $C \equiv !(A \sqcap B) \star !(A \sqcap B)$. The commutativity of the outer rectangle is the desired identity. We leave the verification of $\epsilon \circ \mathbf{p}_{A,B} = \mathbf{e}_{A \sqcap B}$ to the reader.

We define

$$
\begin{aligned}
\mathbf{k}_1 &:= \mathbf{r}_A \circ \rho \circ (\mathrm{id}_{!A} \star \mathbf{e}_B), \\
\mathbf{k}_2 &:= \mathbf{r}_B \circ \lambda \circ (\mathbf{e}_A \star \mathrm{id}_{!B}), \\
\mathbf{k} &= \langle \mathbf{k}_1, \mathbf{k}_2 \rangle, \\
\mathbf{q}_{A,B} &:= g(\mathbf{k}, \lambda \circ (\mathbf{e}_A \star \mathbf{e}_B), m \circ (\mathbf{d}_A \star \mathbf{d}_B)) \equiv g(\mathbf{k}).
\end{aligned}
$$

We show that $\mathbf{q} = \mathbf{p}^{-1}$, as follows.

$$
g(\mathbf{k}) \circ \mathbf{p} = g(\mathbf{k} \circ \mathbf{p}) = (\text{ Lemma 12.8 }) g(\langle \mathbf{k}_1, \mathbf{k}_2 \rangle \circ \mathbf{p}) = \\
g(\langle \mathbf{k}_1 \circ \mathbf{p}, \mathbf{k}_2 \circ \mathbf{p} \rangle) = g(\langle \mathbf{h}_1, \mathbf{h}_2 \rangle)
$$

where by definition

$$
\begin{aligned}
\mathbf{h}_1 &:= \mathbf{r}_A \circ \rho \circ (\mathrm{id}_{!A} \star \mathbf{e}_B) \circ (!\pi_0 \star !\pi_1) \circ \mathbf{d}_{A \sqcap B}, \\
\mathbf{h}_2 &:= \mathbf{r}_B \circ \lambda \circ (\mathbf{e}_A \star \mathrm{id}_{!B}) \circ (!\pi_0 \star !\pi_1) \circ \mathbf{d}_{A \sqcap B}.
\end{aligned}
$$

Then

$$
\begin{aligned}
\mathbf{h}_1 &:= \mathbf{r}_A \circ \rho \circ (\mathrm{id}_{!A} \star \mathbf{e}_A) \circ (!\pi_0 \star !\pi_0) \circ \mathbf{d}_{A \sqcap B}, \\
\mathbf{h}_2 &:= \mathbf{r}_B \circ \lambda \circ (\mathbf{e}_B \star \mathrm{id}_{!B}) \circ (!\pi_1 \star !\pi_1) \circ \mathbf{d}_{A \sqcap B}
\end{aligned}
$$

since $\mathbf{e}_B \circ !\pi_1 = \mathbf{e}_{A \sqcap B} = \mathbf{e}_A \circ !\pi_0$, and hence

$$
\begin{aligned}
\mathbf{h}_1 &:= \mathbf{r}_A \circ \rho \circ (\mathrm{id}_{!A} \star \mathbf{e}_A) \circ \mathbf{d}_A \circ !\pi_0, \\
\mathbf{h}_2 &:= \mathbf{r}_B \circ \lambda \circ (\mathbf{e}_B \star \mathrm{id}_{!B}) \circ \mathbf{d}_B \circ !\pi_1
\end{aligned}
$$

since $!\pi_0$ and $!\pi_1$ are comonoid morphisms. Therefore

$$g(\langle \mathbf{h}_1, \mathbf{h}_2 \rangle) = g(\langle \mathbf{r}_A \circ !\pi_0, \mathbf{r}_B \circ !\pi_1 \rangle)$$

by properties of comonoids, and so

$$g(\langle \mathbf{h}_1, \mathbf{h}_2 \rangle) = g(\langle \pi_0 \circ \mathbf{r}_{A \cap B}, \pi_1 \circ \mathbf{r}_{A \cap B} \rangle) = g(\mathbf{r}_{A \cap B}) = \mathrm{id}_{!(A \cap B)}$$

(naturality of $\mathbf{r}$). We continue with the other half of the verification:

$$
\begin{aligned}
\mathbf{p} \circ g(\mathbf{k}) &= (!\pi_0 \star !\pi_1) \circ \mathbf{d}_{A \cap B} \circ g(\mathbf{k}) \\
&= (!\pi_0 \star !\pi_1) \circ (g(\mathbf{k}) \star g(\mathbf{k})) \circ \delta \, (g \text{ comonoid morphism}) \\
&= (!\pi_0 \star g(\mathbf{k})) \star (!\pi_1 \star g(\mathbf{k})) \circ \delta \\
&= g(\pi_0 \circ \mathbf{k}) \star g(\pi_1 \star \mathbf{k}) \circ \delta \, (\text{Lemma 12.8}) \\
&= (g(\mathbf{k}_1) \star g(\mathbf{k}_2)) \circ \delta \, (\text{definition of } \mathbf{k}) \\
&= (g(\mathbf{r}_A \circ \rho \circ (\mathrm{id}_{!A} \star \mathbf{e}_B)) \star (g(\mathbf{r}_B \circ \lambda \circ (\mathbf{e}_A \star \mathrm{id}_{!B})) \circ \delta \\
&= (\rho \circ (\mathrm{id}_{!A} \star \mathbf{e}_B)) \star (\lambda \circ (\mathbf{e}_A \star \mathrm{id}_{!B})) \circ \delta
\end{aligned}
$$

(since $g(\mathbf{r}_A \circ f) = g(\mathbf{r}_A) \circ f = f$ etc.), hence

$$
\begin{aligned}
\mathbf{p} \circ g(\mathbf{k}) &= (\rho \circ (\mathrm{id}_{!A} \star \mathbf{e}_B)) \star (\lambda \circ (\mathbf{e}_A \star \mathrm{id}_{!B})) \circ m \circ (\mathbf{d}_A \star \mathbf{d}_B) \\
&\qquad\qquad\qquad\qquad\qquad\qquad\qquad (\text{definition of } \mathbf{d}) \\
&= (\rho \circ (\mathrm{id}_{!A} \star \mathbf{e}_A)) \star (\lambda \circ (\mathbf{e}_B \star \mathrm{id}_{!B})) \circ (\mathbf{d}_A \star \mathbf{d}_B) \\
&\qquad\qquad\qquad\qquad\qquad\qquad\qquad (\text{application of } m) \\
&= (\rho \circ (\mathrm{id}_{!A} \star \mathbf{e}_A) \circ \mathbf{d}_A) \star (\lambda \circ (\mathbf{e}_B \star \mathrm{id}_{!B}) \circ \mathbf{d}_B)) \\
&= \mathrm{id}_{!A} \star \mathrm{id}_{!B} = \mathrm{id}_{!A \star !B}.
\end{aligned}
$$

The naturality of the construction is obvious. $\square$

♠ EXERCISE. Show that $\mathbf{e}_\top$ has an inverse $g(\top_\mathbf{1}, \lambda^{-1}, \mathbf{e}_\top)$; hence we can take $\mathbf{t} := \mathbf{e}_\top$ in the theorem.

12.11. Purely logically, the addition of free storage to an ILC corresponds to a rule of *free storage*

FS
$$\frac{A \Rightarrow 1 \quad A \Rightarrow A \star A \quad A \Rightarrow B}{A \Rightarrow !B}$$

in addition to the rules

$$\frac{\Gamma \Rightarrow C}{\Gamma, !A \Rightarrow C} \qquad \frac{\Gamma, A \Rightarrow C}{\Gamma, !A \Rightarrow C} \qquad \frac{\Gamma, !A, !A \Rightarrow C}{\Gamma, !A \Rightarrow C}$$

FS may also be formulated as

$$\frac{A \Rightarrow 1 \sqcap (A \star A) \sqcap B}{A \Rightarrow !B}$$

Clearly, Theorem 9.3 may be extended to an equivalence between ILL_e on the one hand and C-ILL_e with the rule

$$\text{CFS} \quad \frac{\epsilon : A \longrightarrow 1 \quad \delta : A \longrightarrow A \star A \quad \phi : A \longrightarrow B}{g(\phi, \epsilon, \delta) : A \longrightarrow !B}$$

(combinatorial free storage) on the other hand.

12.12. Proposition. *A complete IL-algebra* $(X, \sqcap, \sqcup, \bot, \multimap, \star_c, 1)$ *becomes an IL-algebra with free storage if one takes*

$$!x = \prod\nolimits_{n \in \mathbb{N}} (1 \sqcap x)^n$$

where $y^0 := 1$, $y^{n+1} := y \star y^n$. *The soundness theorem w.r.t. interpretations in IL-algebras extends to a soundness theorem for* $\text{CLL}_e +$ *FS w.r.t. interpretations in IL-algebras with free storage as defined.*

♠ Exercise. Prove the proposition.

Corollary. $\text{CLL}_e +$ *FS is conservative over* CLL_0.
Proof. Immediate, since we can extend any IL-algebra which is complete as a lattice to an IL-algebra with free storage; this applies therefore to the canonical phase structure of 8.10, which validates exactly the sequents of CLL_0. □

12.13. Proposition. Lin, *the category of Girard domains with linear maps, becomes a CLC with free storage, if one takes for* $!(A, \sim_A)$ *the set* $(X, \sim)$ *where* X *is inductively generated by the clauses*

$$(0, 0) \in X;$$
$$a \in A \Rightarrow (1, a) \in X;$$
$$a, b \in X \Rightarrow (2, (a, b)) \in X;$$

and $\sim$ *on* X *is inductively generated by*

$$\sim \text{ is symmetric and reflexive;}$$
$$a \sim_A b \Rightarrow (1, a) \sim (1, b);$$
$$a \sim a', \ b \sim b' \Rightarrow (2, (a, b)) \sim (2, (a', b')).$$

♠ Exercise. Show that $!A$ as defined is a free comonoid in Lin, and the least fixed point of $X \cong 1 \sqcap A \sqcap (X \star X)$.

13 Evaluation in typed calculi

In this chapter we shall study some computational aspects of intuitionistic propositional logic and of $\mathbf{ILL_e}$, seen as type theories. The case of intuitionistic logic serves as introduction to the study of $\mathbf{ILL_e}$. For our exposition we made use of Abramsky 1990.

13.1. Evaluation of closed terms for intuitionistic logic

The idea of evaluation to be explained becomes clearer if we do not restrict ourselves to the "pure" calculus of terms of intuitionistic logic, but add constants for the elements of primitive types (proposition letters); say $\{\bar{p}_{i,j} : j \in J_{P_i}\}$ is the set of constants of type P_i. For arbitrary prime propositions P, Q we use $\bar{p}, \bar{p}', \ldots : P$, $\bar{q}, \bar{q}', \ldots : Q$ for such constants. In addition we may assume that there are certain function symbols $\phi, \phi', \ldots$ denoting functions between primitive types. To fix ideas, let us take as example a single function symbol ϕ denoting a total function f mapping the objects of type P to objects of type Q; we shall assume that for each $\bar{p} : P$ there is a constant $\bar{q} \equiv \overline{fp}$, $\bar{q} : Q$.

If we regard the primitive types as representing the kind of data we are interested in, and the higher types as computational auxiliaries, the principal aim in evaluating terms is that the closed terms of primitive types can be reduced to constants. As to the more complex types, we have some freedom. In order to account for the constants and ϕ, we have to add to the calculus axioms

$$\bar{p}_{i,j} : P_i \qquad \frac{\bar{p} : P}{\phi \bar{p} : Q}$$

For the evaluation of closed terms, we can choose a "lazy evaluation" strategy, or an "eager evaluation" strategy (or a mixture of the two). In both cases we are not trying to bring all terms to normal form (a term is normal if no conversion rule to any subterm applies, i.e., the term contains no redexes), but only in bringing the terms in "canonical form". In lazy evaluation, we evaluate "from the outside" and stop as soon as a term is recognizable as having been generated by an introduction rule for the appropriate type. Lazy evaluation corresponds to "call-by-name" of programming; as long as possible we leave (sub)terms unevaluated, so as to avoid the labour of evaluating a term which afterwards turns out to be not needed.

In the case of eager evaluation, corresponding to "call-by-value" in programming, we evaluate from within, we work as much as possible with terms evaluated before.

In the case of *lazy* evaluation, the (*lazy*) *canonical terms* are

$$\bar{p}, 1, \lambda x.t, \langle s, t \rangle, \kappa_0 t, \kappa_1 t$$

(s, t terms). So if we have arrived at these forms we are done with evaluating. For *eager* evaluation, we have (*eager*) *canonical terms*

$$\bar{p}, 1, \lambda x.t, \langle c, d \rangle, \kappa_0 c, \kappa_1 c$$

(c, d canonical, t arbitrary). We do not require t in $\lambda x.t$ to be canonical, since we consider evaluation of closed terms only. (If we also wanted to evaluate terms with variables, we might take $\lambda x.c$ as canonical, as well as variables. Then we obtain the usual notion of normal form.)

The reduction relation $\triangleright$ for lazy evaluation is inductively specified by (c canonical, t, t_i, s_i, s' arbitrary terms, $i \in \{0, 1\}$):

$$c \triangleright c \qquad \phi \bar{p} \triangleright \overline{fp} \qquad \frac{t \triangleright \langle s_0, s_1 \rangle \quad s_i \triangleright c}{\pi_i t \triangleright c}$$

$$\frac{t \triangleright \lambda x.t' \quad t'[x/s] \triangleright c}{ts \triangleright c} \qquad \frac{t \triangleright \kappa_i s \quad t_i[x_i/s] \triangleright c}{\mathrm{E}^{\vee}_{x_0, x_1}(t, t_0, t_1) \triangleright c}$$

For eager evaluation we have

$$c \triangleright c \qquad \phi \bar{p} \triangleright \overline{fp}$$

$$\frac{t \rhd c \quad s \rhd d}{\langle t,s \rangle \rhd \langle c,d \rangle} \qquad \frac{t \rhd \langle c_0,c_1 \rangle}{\pi_i t \rhd c_i}$$

$$\frac{t \rhd \lambda x.t' \quad s \rhd c \quad t'[x/s] \rhd d}{ts \rhd d}$$

$$\frac{t \rhd c}{\kappa_i t \rhd \kappa_i c} \qquad \frac{t \rhd \kappa_i c \quad t_i[x_i/c] \rhd d}{E^{\vee}_{x_0,x_1}(t,t_0,t_1) \rhd d}$$

In consequence of the strong normalization theorems for intuitionistic finite type theory and propositional logic (see e.g., Troelstra 1973, Section 4.1) we are certain in this case that both strategies always lead to canonical forms. (it is also possible to mix lazy and eager strategies: lazy for some types, eager for others)

PROPOSITION. *Lazy and eager evaluation are both determinate, i.e., if $t \rhd c$ and $t \rhd d$ for canonical forms c,d, then $c \equiv d$, modulo the renaming of bound variables.*

PROOF. The proof is by induction on the length of deductions of $t \rhd c$. Note that for each t exactly one rule applies for obtaining $t \rhd c$, with the exception of the rule for $E^{\vee}$; but in this case there is a shorter deduction of $t \rhd \kappa_i s$ (lazy) or $t \rhd \kappa_i c$ (eager) for precisely one i (by induction hypothesis). $\square$

13.2. Evaluation of closed terms in N-ILL$_e$

In this case we choose an evaluation strategy which is a mixture of eager and lazy evaluation.

For terms s of tensor type, if an elimination constant $E^{\star}(s,t)$ is applied, s is always a (tensor-)pair of which both components are used, so that eager evaluation seems advisable. Similarly, if a term s : $A \multimap B$ is of the form $\lambda x.s'$, then s' contains x exactly once; so when it is used in a β-conversion applied to $(\lambda x.s')(t)$ it seems advantageous to have t evaluated beforehand (hence again eager evaluation). For a term $t : A \sqcup B$, in order to use it in a conversion we need it in a form $\kappa_0 t$ or $\kappa_1 t$; this again suggests eager evaluation.

But if $t : A \sqcap B$ is used in a computation step, we have $t \equiv \langle t_0,t_1 \rangle$ and π_0, π_1 is applied, i.e., we need only one of the components; here lazy evaluation seems better. Similarly for t : $!A$, because of the weakening rule where t is not actually used, lazy evaluation seems preferable. Note that writing intuitionistic implication $A \to B$ as $!A \multimap B$ leads to a lazy evaluation for $A \to B$, because of the lazy evaluation for $!A$.

We refrain from adding constants for the elements of primitive types, and consider the pure calculus only. In our evaluation we now use as *canonical* terms:

$$1, \langle t, s \rangle, !t, *, c \star d, \lambda x.t, \kappa_0 c, \kappa_1 c$$

(c, d canonical, t, s arbitrary).
Evaluation
(c, d, e canonical, t, t_i, t', s arbitrary terms, $i \in \{0, 1\}$).

$$c \triangleright c \qquad \frac{t \triangleright * \quad s \triangleright c}{\mathrm{E}^{\mathbf{1}}(t, s) \triangleright c}$$

$$\frac{s \triangleright c \quad t \triangleright d}{s \star t \triangleright c \star d} \qquad \frac{s \triangleright c \star d \quad t[x, y/c, d] \triangleright e}{\mathrm{E}^{\star}_{x,y}(s, t) \triangleright e}$$

$$\frac{t \triangleright \lambda x.t' \quad s \triangleright c \quad t'[x/c] \triangleright d}{ts \triangleright d} \qquad \frac{t \triangleright \langle t_0, t_1 \rangle \quad t_i \triangleright c}{\pi_i t \triangleright c}$$

$$\frac{t \triangleright c}{\kappa_i t \triangleright \kappa_i c} \qquad \frac{t \triangleright \kappa_i c \quad t_i[x_i/c] \triangleright d}{\mathrm{E}^{\sqcup}_{x_0,x_1}(t, t_0, t_1) \triangleright d}$$

$$\frac{t \triangleright !t' \quad t' \triangleright c \quad s[x/c] \triangleright d}{\mathrm{E}^{!}_{x}(t, s) \triangleright d} \qquad \frac{s \triangleright c}{\mathrm{E}^{\mathrm{w}}(t, s) \triangleright c} \qquad \frac{s[x, y/t, t] \triangleright c}{\mathrm{E}^{\mathrm{c}}_{x,y}(t, s) \triangleright c}$$

N.B. Type $\top$ is computationally uninteresting; also $\mathrm{E}^{\perp}$ does not play a role, since the obvious clause "$\mathrm{E}^{\perp}(c)$ is canonical for canonical c" remains empty because there is no closed canonical $c : \perp$, as is readily seen by interpreting all logical operators by their intuitionistic analogues. As before we have

PROPOSITION. *Evaluation is determinate.*

13.3.   We can also prove that evaluation is a total function on closed terms. This can be done by a method which is a simplification (due to Abramsky (1990)) of Girard's original proof of strong normalization for terms in the "system F" (cf. Girard, Lafont and Taylor 1988). We now treat the term calculus as a Curry type-assignment system, i.e., the terms can be typed, but do not carry types themselves.

We regard the propositional variables $P, P', P'', \ldots, Q, Q', \ldots$ as type variables and add universal type quantifiers (propositional quantifiers) $\forall P$ to the system, with the rules

$$\forall \mathrm{I} \; \frac{\Gamma \Rightarrow t : A(P)}{\Gamma \Rightarrow t : \forall P \, A(P)} \qquad \forall \mathrm{E} \; \frac{\Gamma \Rightarrow t : \forall P \, A}{\Gamma \Rightarrow t : A[P/B]}$$

(P not free in Γ, B free for P in A). Thus the terms assigned do not depend on the type variables.

DEFINITION. A *computability type* X is a set of closed terms which have a canonical form also in X, and such that if $c \in X$, $t \vartriangleright c$, then $t \in X$. We define the sets $!, \top, \bot$ and the operators $\sqcap, \sqcup, \star, \multimap, !$ on computability types by (c, d canonical):

$$\top := \{1\};$$
$$\bot := \emptyset;$$
$$1 := \{t : t \vartriangleright *\};$$
$$X \star Y := \{t : t \vartriangleright c \star d, c \in X, d \in Y\};$$
$$X \multimap Y := \{t : t \vartriangleright \lambda x.t', \forall s \in X(ts \in Y)\};$$
$$X \sqcap Y := \{t : t \vartriangleright \langle s, s' \rangle, s \in X, s' \in Y\};$$
$$X \sqcup Y := \{t : (t \vartriangleright \kappa_0 c \text{ and } c \in X) \text{ or } (t \vartriangleleft \kappa_1 c \text{ and } c \in Y)\};$$
$$!X := \{t : t \vartriangleright !s, s \in X\}.$$

and, if F is a mapping from computability types to computability types, we put

$$\forall(F) := \cap\{F(X) : X \text{ computability type }\}.$$

Given an assignment ρ of computability types to the propositional variables, we can assign a set of terms $[\![A]\!]_\rho$ to each second-order propositional formula (second-order type) A, where $[\![A \diamond B]\!]_\rho = [\![A]\!]_\rho \diamond [\![B]\!]_\rho$ for $\diamond \in \{\star, \multimap, \sqcap, \sqcup\}$, $[\![\forall P\, A]\!]_\rho = \forall(\lambda X.[\![A]\!]_{\rho[P/X]})$ where $\rho[P/X](Q) = \rho(Q)$ for $Q \neq P$, $\rho(P) = X$. We define

$$(\vec{x} : \Gamma \models s : A) := \forall\rho\forall\vec{t} \in [\![\Gamma]\!]_\rho(s[\vec{x}/\vec{t}] \in [\![A]\!]_\rho)$$

Here the Γ, Δ etc are multisets of formulas again, not of typing statements; we shall use obvious abbreviations in writing $\vec{t} : \Gamma$, $\vec{t} \in [\![\Gamma]\!]_\rho$, etc. $\square$

LEMMA. *For all assignments ρ of computability types and all formulas A, $[\![A]\!]_\rho$ is a computability type.*
PROOF. By induction on the complexity of A, using the fact that $t \vartriangleright c$ is always established by a uniquely determined rule application. $\square$

PROPOSITION. *If $\Gamma \Rightarrow t : A$ is derivable in N-**ILL**$_e$ with terms, then $\Gamma \models t : A$.*

PROOF. By induction on the length of derivations of $\Gamma \Rightarrow t : A$. We check some typical cases.

Case 1. Let the final rule applied be an instance of $\multimap$ E:

$$\frac{\vec{x} : \Gamma \Rightarrow t : A \multimap B \qquad \vec{y} : \Delta \Rightarrow s : A}{\vec{x}, \vec{y} : \Gamma, \Delta \Rightarrow ts : B}$$

We have to show for given ρ, and $\vec{u}, \vec{v} \in [\![\Gamma, \Delta]\!]_\rho$, that $t[\vec{x}, \vec{y}/\vec{u}, \vec{v}] \in [\![B]\!]_\rho$. So assume, $\vec{u} \in [\![\Gamma]\!]_\rho$, $\vec{v} \in [\![\Delta]\!]_\rho$; by induction hypothesis

$$t' \equiv t[\vec{x}/\vec{u}] \in [\![A \multimap B]\!]_\rho, \quad s' \equiv s[\vec{y}/\vec{v}] \in [\![A]\!]_\rho$$

Then $t' \rhd \lambda x.t''$, $\forall s'' \in [\![A]\!]_\rho (t's'' \in [\![B]\!]_\rho)$, so $t's' \in [\![B]\!]_\rho$.

Case 2. For notational simplicity we drop side formulas and ρ; let the last rule applied be an instance of $\sqcup$E:

$$\frac{\Rightarrow t : A \sqcup B \qquad z' : A \Rightarrow s' : C \qquad z'' : B \Rightarrow s'' : C}{\Rightarrow \mathrm{E}^\sqcup_{z',z''}(t, s', s'') : C}$$

By the induction hypothesis we have from the first premise a $c \in [\![A]\!]$, $t \rhd \kappa_0 c$ or $c \in [\![B]\!]$, $t \rhd \kappa_1 c$, say the first, and also from the induction hypothesis applied to the second premise:

$$t' \in [\![A]\!] \Rightarrow s'[z'/t'] \in [\![C]\!],$$

Hence in the first case $s'[z'/c] \rhd c \in [\![C]\!]$. Then $\mathrm{E}^\sqcup_{z',z''}(t, s', s'') \rhd c \in [\![C]\!]$. Similarly if $t \rhd \kappa_1 c$.

Case 3. Let the last rule application be

$$\frac{\Rightarrow t : {!B} \qquad x : B \Rightarrow s : A}{\mathrm{E}^!_x(t, s) : A}$$

By the induction hypothesis for the first premise $t \rhd {!t'}$, $t' \in [\![B]\!]$. The second hypothesis yields that for all $t'' \in [\![B]\!]$, $s[x/t''] \in [\![A]\!]$, hence $s[x/t'] \rhd d \in [\![A]\!]$ for some canonical d. It follows that $\mathrm{E}^!_x(t, s) \rhd d$, so $\mathrm{E}^!_x(t, s) \in [\![A]\!]$.

Case 5. Let $\Rightarrow t : A(P)$. Obviously t cannot depend on a variable $x : P$, so $t \in [\![A(P)]\!]_\rho$ gives also $\forall X (t \in [\![A(P)]\!]_{\rho[P/X]})$, so $t \in [\![\forall P\, A(P)]\!]_\rho$. The other cases are left to the reader.$\square$

13.4. COROLLARY. *For all closed t such that $\Rightarrow t : A$ there is a canonical c such that $t \rhd c$.*

PROOF. Assume $\Rightarrow t : A$. Without loss of generality we may assume A to be closed (simply apply $\forall$I a few times if necessary). Then it follows that $t \models A$, i.e., $t \in [\![A]\!]$. But $[\![A]\!]$ is a computability type, and all terms in a computability type have a canonical form. $\square$

14 Computation by lazy evaluation in CCC's

A straightforward implementation of computations based on the evaluation mechanism of the preceding chapter would call for a type of abstract machine capable of handling bound variables and the assignment of values to free variables (cf. the SECD-machines in Abramsky 1990). The handling of bound variables may be avoided by developing evaluation mechanisms based on the combinatorial calculi introduced in Chapter 9, where all terms denote arrows in categories. First we study the case of CCC's (for background see for example Curien 1986). The next chapter gives a similar treatment for intuitionistic linear categories. In Chapter 16 we show how to implement the abstract evaluations of the preceding two chapters by (abstract) machines. This and the next two chapters are based on Lafont 1988a, 1989.

14.1. Embedding a category in a CCC

As we have seen in Chapter 9, it is easy to construct *free* cartesian closed, monoidal, symmetric monoidal, intuitionistic linear etc. categories over a set of generators, since all these types of categories are given by sets of equations for the arrows, and this construction readily generalizes to the construction of free categories over a directed graph G.

In particular, G itself may be a category $\mathcal{C}$, in which case we require that composition in the free category extends composition in $\mathcal{C}$ etc.

Now for any type of category (say X-category) the following question arises. Let $X(\mathcal{C})$ be the free X-category constructed over the basis category $\mathcal{C}$. There is an obvious embedding functor $J : \mathcal{C} \longrightarrow X(\mathcal{C})$ with $J(A) := A$ for objects, and $J(f) := (f/ \equiv) : A \longrightarrow B$ for morphisms $f : A \longrightarrow B$ in $\mathcal{C}$; here $\equiv$ is the equivalence relation generated by the identities for X-categories. We may now ask: is the extension of $\mathcal{C}$ to $X(\mathcal{C})$ *conservative*, or in other words, is J *full* (no new arrows in $X(\mathcal{C})$ between objects of $\mathcal{C}$) and *faithful* (no distinct arrows of $\mathcal{C}$ are identified in $X(\mathcal{C})$).

As we shall see, both for CCC's and for intuitionistic linear categories J is indeed full and faithful.

Logically, it may be seen as a weak version of the subformula principle: for atomic formulas P,Q (objects in $\mathcal{C}$) $P \Rightarrow Q$ is provable in the combinatorial version of intuitionistic logic only if it is an axiom (represented by an arrow $f : P \longrightarrow Q$ in $\mathcal{C}$). Actually, more is true: every term for an arrow between atomic formulas reduces to an arrow in $\mathcal{C}$, so the existing axioms also do not get new proofs in $X(\mathcal{C})$ (modulo the equivalence relation on arrows in $X(\mathcal{C})$).

However, the most interesting aspect of the proof of conservativeness in this chapter is the use of abstract values, since that method will suggest the method of implementation in Chapter 16.

14.2. Faithfulness of J for CCC's

For any category $\mathcal{C}$, let $\mathcal{C}^*$ be the free CCC over $\mathcal{C}$. By a simple category-theoretical argument we can prove

PROPOSITION. $J : \mathcal{C} \longrightarrow \mathcal{C}^*$ *is faithful.*

PROOF. It suffices to find *some* CCC $\mathcal{D}$ into which $\mathcal{C}$ can be faithfully embedded by a functor F; if F does not identify arrows from $\mathcal{C}$, then a fortiori J does not do so. In other words, the result follows, since we have a factorization because of the free character of $\mathcal{C}^*$.

$$
\begin{array}{ccc}
\mathcal{C} & \xrightarrow{\ \ F\ \ } & \mathcal{D} \\
\downarrow{\scriptstyle J} & \circledast & \uparrow{\scriptstyle G} \\
\mathcal{C}^* & = & \mathcal{C}^*
\end{array}
$$

(G may possibly identify arrows from $\mathcal{C}^*$). For $\mathcal{D}$ we can take the presheaf category $\mathsf{Set}^{\mathcal{C}^{\mathrm{op}}}$, and as faithful embedding the Yoneda-functor Y:

$$\begin{cases} Y(D) := \mathcal{C}(-, D) \\ Y(f) := \mathrm{Nat}(\mathcal{C}(-, D), \mathcal{C}(-, D')) \simeq \mathcal{C}(D, D') \text{ for } f : D \longrightarrow D' \end{cases}$$

The Yoneda lemma tells us that Y is faithful; it remains to show that $\mathcal{D} \equiv \mathsf{Set}^{\mathcal{C}^{\mathrm{op}}}$ is a CCC, which is a standard result from category theory (see e.g., Lambek & Scott 1986, II.9). $\square$

14.3. COROLLARY. *Let* $\mathrm{ILC}(\mathcal{C})$ *be the free intuitionistic linear category constructed over* $\mathcal{C}$; *the embedding* $J : \mathcal{C} \longrightarrow \mathrm{ILC}(\mathcal{C})$ *is faithful.*
PROOF. Immediate, since $\mathcal{C}^*$ identifies only *more* arrows than $\mathrm{ILC}(\mathcal{C})$. $\square$

14.4. The fullness of the embedding for CCC's

$J : \mathcal{C} \longrightarrow \mathcal{C}^*$ is also full, and this fact can be proved by a categorical argument (Lafont 1989), which we shall not present here. However, as we shall show later, $J' : \mathcal{C} \longrightarrow \mathrm{ILC}(\mathcal{C})$ is full, and hence a fortiori $J : \mathcal{C} \longrightarrow \mathcal{C}^*$ is full, and this fact will be proved using a notion of evaluation.

We demonstrate this method first by establishing fullness of J for a very simple case, namely where $\mathcal{C}$ is a category with one object X and a single arrow id_X. It is instructive, in following the developments below, to think of X as a singleton set and arrows $X \longrightarrow A$ as elements of A.

NOTATION. $\mathrm{Comb}(A, B)$ is the set of all *terms* denoting arrows (i.e., combinator terms) in $\mathcal{C}^*(A, B)$. $\square$

To each object $A \in \mathcal{C}^*$ we assign a set of abstract values $\mathrm{Val}(A)$, and a representation function

$$\ulcorner \, \urcorner : \mathrm{Val}(A) \longrightarrow \mathrm{Comb}(X, A),$$

and to each combinator term $\phi : A \longrightarrow B$ we assign a function $|\phi| : \mathrm{Val}(A) \longrightarrow \mathrm{Val}(B)$ (the *semantical interpretation* of ϕ) which is compatible with $\ulcorner \, \urcorner$, that is to say

compat $\qquad\qquad \ulcorner(|\phi|u)\urcorner \equiv \phi \circ \ulcorner u \urcorner$

or diagrammatically

$$
\begin{array}{ccc}
X & = & X \\
\downarrow^{\ulcorner u \urcorner} \quad \circledast & & \downarrow^{\ulcorner (|\phi|u) \urcorner} \\
A & \xrightarrow{\ \phi\ } & B
\end{array}
$$

As equations for a CCC we take the set specified in Table 11.

14.5. DEFINITION. (Val and $\ulcorner\ \urcorner$)

(i) $\mathrm{Val}(X) := \{*\}$, $\mathrm{Val}(\top) := \{t\}$;
$\ulcorner * \urcorner := \mathrm{id}_X : X \to X$, $\ulcorner t \urcorner := \top_X$.

(ii) $\mathrm{Val}(A \wedge B) := \mathrm{Val}(A) \times \mathrm{Val}(B)$;
$\ulcorner (u, v) \urcorner := \langle \ulcorner u \urcorner, \ulcorner v \urcorner \rangle : X \to A \wedge B$.

(iii) $\mathrm{Val}(A \to B) := \{(\phi, f) : \phi \in \mathrm{Comb}(X \wedge A, B)$,
$f : \mathrm{Val}(A) \to \mathrm{Val}(B)$, $\forall u (\ulcorner fu \urcorner = \phi \circ \langle \mathrm{id}, \ulcorner u \urcorner \rangle)\}$;
$\ulcorner (\phi, f) \urcorner := \mathrm{cur}(\phi) : X \to (A \to B)$.

So a value of $A \to B$ contains a syntactical and a semantical component satisfying together a compatibility condition. $\square$

14.6. DEFINITION. (inductive definition of $|\ |$)

(i) $|\mathrm{id}|u := u$, $|\top_A|u := t$,

(ii) $|\phi \circ \psi|u := |\phi|(|\psi|u)$,

(iii) $|\pi_0|(u, v) := u$,

(iv) $|\pi_1|(u, v) := v$,

(v) $|\langle \phi, \psi \rangle|u := (|\phi|u, |\psi|v)$,

(vi) $|\mathrm{ev}|((\phi, f), u) := fu$,

(vii) $|\mathrm{cur}(\phi)|u := (\phi \circ \langle \ulcorner u \urcorner \circ \pi_0, \pi_1 \rangle, fu)$, where $fuv = |\phi|(u, v)$. $\square$

14.7. LEMMA. *For all combinator terms ϕ*

$$\ulcorner (|\phi|u) \urcorner = \phi \circ \ulcorner u \urcorner.$$

PROOF.

Case (i) $\ulcorner (|\mathrm{id}|u) \urcorner \equiv \ulcorner u \urcorner \equiv \mathrm{id} \circ \ulcorner u \urcorner (idl)$, etc.

Case (ii) $\ulcorner(|\phi \circ \psi|u)\urcorner \equiv \ulcorner(|\phi|(\psi|u))\urcorner \equiv \phi \circ \ulcorner(|\psi|u)\urcorner \equiv \phi \circ (\psi \circ \ulcorner u \urcorner) \equiv (\phi \circ \psi) \circ \ulcorner u \urcorner$ (ass).

Case (iii) $\ulcorner(|\pi_0|(u,v))\urcorner = \ulcorner u \urcorner \equiv \pi_0 \circ \langle \ulcorner u \urcorner, \ulcorner v \urcorner \rangle \equiv \pi_0 \circ \ulcorner(u,v)\urcorner$ (prl).

Case (iv) Similarly.

Case (v) $\ulcorner(|\langle \phi, \psi \rangle|u)\urcorner = \ulcorner(|\phi|u, |\psi|u)\urcorner = \langle \ulcorner(|\phi|u)\urcorner, \ulcorner(|\psi|u)\urcorner \rangle = \langle \phi \circ \ulcorner u \urcorner, \psi \circ \ulcorner u \urcorner \rangle \equiv \langle \phi, \psi \rangle \circ \ulcorner u \urcorner$ (pair).

Case (vi) $\ulcorner(|\mathrm{ev}|((\phi, f), u))\urcorner = \ulcorner(fu)\urcorner = \phi \circ \langle \mathrm{id}, \ulcorner u \urcorner \rangle = \mathrm{ev} \circ \langle \mathrm{cur}(\phi), \ulcorner u \urcorner \rangle (ev) = \mathrm{ev} \circ \langle \ulcorner(\phi, f)\urcorner, \ulcorner u \urcorner \rangle = \mathrm{ev} \circ \ulcorner((\phi, f), u)\urcorner$.

Case (vii) We have to check $|\mathrm{cur}(\phi)|u$ is a value. If $\phi : A \wedge B \longrightarrow C$, then $\mathrm{cur}(\phi) : A \longrightarrow (B \to C)$, $|\mathrm{cur}(\phi)| : \mathrm{Val}(A) \longrightarrow \mathrm{Val}(B \to C)$. So if $u \in \mathrm{Val}(A)$ we must show that $|\mathrm{cur}(\phi)|u \in \mathrm{Val}(B \to C)$, i.e.,

$$(\phi \circ \langle \ulcorner u \urcorner \circ \pi_0, \pi_1 \rangle, \lambda v.|\phi|(u, v)) \in \mathrm{Val}(B \to C)$$

For this we have to check that for all $v \in \mathrm{Val}(B)$

$$\ulcorner((\lambda v.|\phi|(u, v))v)\urcorner = \ulcorner(|\phi|(u, v))\urcorner \equiv \phi \circ \langle \ulcorner u \urcorner \circ \pi_0,$$
$$\pi_1 \rangle \circ \langle \mathrm{id}, \ulcorner v \urcorner \rangle = \phi \circ \langle \ulcorner u \urcorner \circ \pi_0 \circ \langle \mathrm{id}, \ulcorner v \urcorner \rangle, \pi_1 \circ \langle \mathrm{id}, \ulcorner v \urcorner \rangle \rangle$$
$$= \phi \circ \langle \ulcorner u \urcorner, \ulcorner v \urcorner \rangle$$
$$= \phi \circ \ulcorner(u, v)\urcorner,$$

which is obviously correct since $\ulcorner(|\phi|(u, v))\urcorner = \phi \circ \ulcorner(u, v)\urcorner$. Next we must show

$$\ulcorner(|\mathrm{cur}|(\phi)u)\urcorner = \ulcorner(\phi \circ \langle \ulcorner u \urcorner \circ \pi_0, \pi_1 \rangle, \lambda v|\phi|(u, v))\urcorner$$
$$= \mathrm{cur}(\phi \circ \langle \ulcorner u \urcorner \circ \pi_0, \pi_1 \rangle) = \mathrm{cur}(\phi) \circ \ulcorner u \urcorner \text{ (cur).} \quad \square$$

14.8. Proof of the fullness of J

Let $\phi \in \mathrm{Comb}(X, X)$. $\mathrm{Val}(X) = \{*\}$, so $|\phi|* = *$, hence

$$\phi \equiv \phi \circ \mathrm{id} = \phi \circ \ulcorner * \urcorner \equiv \ulcorner(|\phi|*)\urcorner = \ulcorner * \urcorner = \mathrm{id}_X. \quad \square$$

Note that not all equations have been used: we did not use pairid and curid.

14.9.   The values of $A \to B$, containing an abstract function, are not concrete, and therefore rather unmanageable. The abstract function plays a role in evaluating $|\mathrm{cur}(\phi)|u$. But if we are only interested in values for objects in $\mathcal{C}$, there is no need to evaluate this expression; in due time "ev" will be applied and the need for evaluating $|\mathrm{cur}(\phi)|u$ disappears. We pursue this idea now by modifying the set-up above; we now assign "lazy values" which are such that expressions need not be evaluated until we arrive at base level.

As values for $A \to B$ we now want to take all (ϕ, u), $\phi \in \mathrm{Comb}(Y \wedge A, B)$, $u \in \mathrm{Val}(Y)$, for *arbitrary* Y. But as it stands this is a circular definition, since $\mathrm{Val}(A \to B)$ now depends on $\mathrm{Val}(Y)$ for arbitrary Y, so in particular on $\mathrm{Val}(A \to B)$ itself.

This problem is solved by defining inductively the typing relation "$u : A$" and the representation $\ulcorner u \urcorner : X \longrightarrow A$ as relations between u and A.

14.10. Definition.   (Lazy values)

(i) $*$ is a value,

(ii) if u, v are values, then also (u, v),

(iii) if ϕ is a combinator and u a value, then $(\phi \cdot u)$ is a value (closure).
$\square$

Definition.   (Typing and $\ulcorner \; \urcorner$)

$$* : X, \quad \ulcorner * \urcorner = \mathrm{id}_X : X \longrightarrow X \qquad t : \top, \quad \ulcorner t \urcorner = \top_X : X \longrightarrow \top$$

$$\frac{u : A \qquad v : B}{(u, v) : A \wedge B} \qquad \frac{u : A \qquad v : B}{\ulcorner (u, v) \urcorner = \langle \ulcorner u \urcorner, \ulcorner v \urcorner \rangle : X \longrightarrow A \wedge B}$$

$$\frac{\phi : A \wedge B \longrightarrow C \qquad u : A}{(\phi \cdot u) : B \to C} \qquad \frac{\phi : A \wedge B \to C \qquad u : A}{\ulcorner (\phi \cdot u) \urcorner = \mathrm{cur}(\phi) \circ \ulcorner u \urcorner : X \to B \to C}$$

$\square$

14.11. Definition.   (Evaluation relation $\triangleright$) We define $\phi u \triangleright v$ (combinator ϕ applied to value u yields value v) inductively by the clauses

$$\frac{u : A}{\mathrm{id}_A u \triangleright u} \qquad \frac{u : A}{\top_A u \triangleright t}$$

$$\frac{\phi : A \longrightarrow B \qquad \psi : B \longrightarrow C \qquad u : A \qquad \phi u \triangleright v : B \qquad \psi v \triangleright w : C}{(\psi \circ \phi) u \triangleright w}$$

$$\frac{u : A \qquad v : B}{\pi_0(u, v) \,\triangleright\, u} \qquad \frac{u : A \qquad v : B}{\pi_1(u, v) \,\triangleright\, v}$$

$$\frac{\phi : A \longrightarrow B \quad \psi : A \longrightarrow C \quad u : A \quad \phi u \,\triangleright\, v : B \quad \psi u \,\triangleright\, w : C}{\langle \phi, \psi \rangle u \,\triangleright\, (v, w)}$$

$$\frac{\phi : A \wedge B \longrightarrow C \quad u : Av : B \quad \phi(u, v) \,\triangleright\, w : C}{\mathrm{ev}((\phi \cdot u), v) \,\triangleright\, w}$$

$$\frac{\phi : A \wedge B \longrightarrow C \quad u : A}{\mathrm{cur}(\phi) u \,\triangleright\, \phi \cdot u}$$

□

REMARK. The treatment of pairing is eager rather than lazy, so the "lazy values" in 14.10 refers to functional types only. For a lazy treatment of pairing, one ought to introduce also a closure for pairing. Then (ii) and (iii) in 14.10 would become: if ϕ, ψ are combinators, u a value, then $\langle \phi, \psi \rangle \cdot u$ and $\mathrm{cur}(\phi) \cdot u$ are values. (We now write $\mathrm{cur}(\phi) \cdot u$ instead of $\phi \cdot u$ since we have to distinguish two types of closure. The typing and evaluation rules for pairing and projections would become:

$$\frac{\phi : A \longrightarrow B \quad \psi : A \longrightarrow C \quad u : A}{\langle \phi, \psi \rangle \cdot u : B \wedge C}$$

$$\frac{\phi : A \longrightarrow B \quad \psi : A \longrightarrow C \quad u : A}{\ulcorner \langle \phi, \psi \rangle u \urcorner = \langle \phi, \psi \rangle \circ \ulcorner u \urcorner : X \longrightarrow B \wedge C}$$

$$\frac{\phi_0 : A \longrightarrow B_0 \quad \phi_1 : A \longrightarrow B_1 \quad u : A}{\langle \phi_0, \phi_1 \rangle u \,\triangleright\, \langle \phi_0, \phi_1 \rangle \cdot u}$$

$$\frac{\phi_0 u \,\triangleright\, v : B_i \quad u : A}{\pi_i(\langle \phi_0, \phi_1 \rangle \cdot u) \,\triangleright\, v}$$

14.12. DEFINITION. Evaluating $\phi : A \longrightarrow B$ on a value $u : A$ means finding a $v : B$ such that $\phi u \,\triangleright\, v$. □

PROPOSITION. *(Uniqueness) For all $\phi : A \longrightarrow B$, $u : A$ there is at most one $v : B$ such that $\phi u \,\triangleright\, v$.*

PROOF. By induction on the derivation of $\phi u \,\triangleright\, v$. □

14.13. PROPOSITION.

$$\phi u \,\triangleright\, v \Rightarrow \ulcorner v \urcorner = \phi \circ \ulcorner u \urcorner.$$

PROOF. We consider two cases and leave the others to the reader as an exercise.

Case (ii) If $(\psi \circ \phi)u \rhd w$ by the second rule, then $((\psi \circ \phi) \circ \ulcorner u \urcorner) = \psi \circ (\phi \circ \ulcorner u \urcorner) = \psi \circ \ulcorner v \urcorner = \ulcorner w \urcorner$.

Case (vi) Let $\mathrm{ev}((\phi \cdot u), v) \rhd w$ by the 6th rule. Then $\mathrm{ev} \circ \ulcorner ((\phi \cdot u), v) \urcorner = \mathrm{ev} \circ \langle \ulcorner (\phi \cdot u) \urcorner, \ulcorner v \urcorner \rangle = \mathrm{ev} \circ \langle \mathrm{cur}(\phi) \circ \ulcorner u \urcorner, \ulcorner v \urcorner \rangle = \mathrm{ev} \circ \langle \mathrm{cur}(\phi \circ \langle \ulcorner u \urcorner \circ \pi_0, \pi_1 \rangle, \ulcorner v \urcorner \rangle$ (cur) $= \phi \circ \langle \ulcorner u \urcorner \pi_0, \pi_1 \rangle \circ \langle \mathrm{id}, \ulcorner v \urcorner \rangle$ (ev) $= \phi \circ \langle \ulcorner u \urcorner, \ulcorner v \urcorner \rangle (pair, prl, prr, id, ass) = \phi \circ \ulcorner (u, v) \urcorner = \ulcorner w \urcorner$ (by the IH applied to the premises). $\square$

Again, note that pairid and curid have not been used.

♠ EXERCISE. Do the remaining cases of the proof.

14.14. THEOREM. *For each combinator* $\phi : A \longrightarrow B$ *and value* $u : A$ *there is a unique value* $v : B$ *such that* $\phi u \rhd v$.
PROOF. For each object A of $\mathcal{C}^*$ we define $[\![A]\!]$, a set of values $u : A$.

(i) $[\![*]\!] := \{*\}$, $[\![\top]\!] := \{t\}$,

(ii) $[\![A \wedge B]\!] := \{(u, v) : u \in [\![A]\!], v \in [\![B]\!]\}$,

(iii) $[\![A \to B]\!] := \{(\phi \cdot u) \; : \; \phi : Y \wedge A \longrightarrow B, \; u : Y,$
   $\forall v \in [\![A]\!] \exists w \in [\![B]\!](\phi(u, v) \rhd w)\}$.

The remainder of the proof is in two lemmas:

LEMMA. *For each combinator* $\phi : A \longrightarrow B$, $u \in [\![A]\!]$, *there is a* $v \in [\![B]\!]$ *such that* $\phi u \rhd v$.
PROOF. By induction on the construction of ϕ. We consider two of the more interesting cases and leave the others as an exercise.

Case(i) $\phi u \equiv \mathrm{ev}(u)$ with $u \in [\![(A \to B) \wedge A]\!]$. Then $u = (v, w)$, $v \in [\![A \to B]\!]$, $w \in [\![A]\!]$; hence $v \equiv (\phi \cdot v') : Y \wedge A \to B$, $v' : Y$ and $\forall w \in [\![A]\!] \exists w' \in [\![B]\!](\phi(v', w) \rhd w')$. Then $\phi(v', w) \rhd w'$ for some $w' \in [\![B]\!]$, hence $\mathrm{ev}((\phi \cdot v'), w) \rhd w'$.

Case(ii) $\phi u \equiv \mathrm{cur}(\psi)u$, with $\psi : A \wedge B \longrightarrow C$, $u \in [\![A]\!]$. Then $\mathrm{cur}(\psi)u \rhd \psi \cdot u \in [\![B \to C]\!]$, (since for $v \in [\![B]\!]$, $\phi(u, v) \rhd w \in [\![C]\!]$ by induction hypothesis). $\square$

LEMMA. *For all* A, $[\![A]\!] = \{u : (u : A)\}$.
PROOF. By induction on the construction of values. The only case which needs to be checked is $A \equiv B \to C$. The only values in $B \to C$ are of the form $(\phi \cdot u)$, $\phi : A \wedge B \longrightarrow C$, $u : A$; then $u \in [\![A]\!]$ (induction

hypothesis), and if $v \in [\![B]\!]$, $(u, v) \in [\![A \wedge B]\!]$, then $\phi(u, v) \,\triangleright\, w$ for some $w \in [\![C]\!]$, hence $(\phi \cdot u) \in [\![B \to C]\!]$. This completes the proof of the theorem. $\Box$

14.15.   *Second proof* of the conservativity theorem.
Let $\phi : X \longrightarrow X$, then $\phi * \,\triangleright*$, so $\ulcorner *\urcorner \equiv \phi \circ \ulcorner *\urcorner$, i.e., $\mathrm{id} \equiv \phi \circ \mathrm{id} \equiv \phi$. $\Box$

REMARK. The verification of (vi) in the proof of the proposition requires

$$\mathrm{ev} \circ \langle \mathrm{cur}(\phi) \circ \ulcorner u \urcorner, \ulcorner v \urcorner \rangle = (\mathrm{cur}') \;\; \phi \circ \langle \ulcorner u \urcorner, \ulcorner v \urcorner \rangle = \phi \circ \ulcorner (u, v) \urcorner$$

where cur' is
$$\mathrm{ev} \circ \langle \mathrm{cur}(\phi) \circ \psi, \chi \rangle = \phi \circ \langle \psi, \chi \rangle.$$

Hence this second proof of fullness relies on idr, ass, prr, prl, pair and cur$'$.

15 Computation by lazy evaluation in SMC's and ILC's

15.1. In this chapter we shall give the proof of the fullness of the embedding functor $J : \mathcal{C} \longrightarrow \mathrm{ILC}(\mathcal{C})$. The strategy is quite similar to the one used for CCC's in the second half of the preceding chapter.

15.2. DEFINITION. For a given directed graph $\mathcal{C}$, we call the nodes *atomic* types, the arrows *atomic* combinators. A tensor product of atomic types is a *primitive* type; a *primitive* combinator is a combinator between primitive types obtained from atomic combinators by composition, tensorproduct and the arrangement combinators α, λ, γ. We use β, β' for primitive combinators. $\square$

In the notion of evaluation to be defined below, the atomic types, or rather their identity arrows, play the role of basic data (not sets of data). The extension from atomic types and atomic combinators to primitive types and combinators is a rather innocent sort of extension. Essentially we introduce nothing but finite groupings of atomic objects and atomic arrows by means of $\star$, and we have complete information concerning the behaviour of standard arrows constructed from α, λ, γ, $\star$ and $\circ$ by the coherence result of Chapter 11. The role of the values is now taken by the canonical combinators, to be defined next.

15.3. Definition. For any type A we define the class $\mathrm{Can}(A)$ of the *canonical combinators* of type A:

(i) If A is atomic, $\mathrm{Can}(A) = \{\mathrm{id}_A\}$.

(ii) $\mathrm{Can}(A \star B) = \{\mu \star \nu : \mu \in \mathrm{Can}(A), \nu \in \mathrm{Can}(B)\}$.

(iii) $\mathrm{Can}(\mathbf{1}) = \{\mathrm{id}_\mathbf{1}\}$.

(iv) For all other types A, $Can(A) = \{(\xi \circ \mu : X \to A) : \xi : Y \to A$ a constructor, $\mu : X \to Y \in Can(Y)\}$. Here a *constructor* is any combinator of the form $\mathrm{cur}(\phi)$, $\top_A$, κ_0, κ_1 or $\langle \phi, \psi \rangle$.

In the next chapter, the canonical combinators will serve as the data on which the programmes act. $\square$

REMARK. This definition does not define $\mathrm{Can}(A)$ by induction on the type-complexity, but as an inductive definition of a relation between μ and A, "$\mu \in \mathrm{Can}(A)$". Thus the last clause (iv) should be read as a rule

$$\frac{\mu : X \longrightarrow Y \in \mathrm{Can}(Y)}{\xi \circ \mu : X \longrightarrow A \in \mathrm{Can}(A)} \quad (\xi : Y \longrightarrow A \text{ a constructor})$$

15.4. Definition. (The evaluation relation $\triangleright$) We define inductively

$$\phi\mu \triangleright \nu; \beta,$$

where

$$\phi : A \longrightarrow B \text{ combinator },$$
$$\mu : X \longrightarrow A, \quad \nu : Y \longrightarrow B \text{ canonical combinators,}$$
$$\beta : X \longrightarrow Y \text{ primitive combinator.}$$

The clauses are

$$\beta \,\mathrm{id} \triangleright \mathrm{id}; \beta \ (\beta \text{ atomic}) \qquad \mathrm{id}\,\mu \triangleright \mu; \mathrm{id}$$

$$\frac{\phi\mu \triangleright \mu'; \beta \qquad \psi\mu' \triangleright \mu''; \beta'}{(\psi \circ \phi)\mu \triangleright \mu''; \beta' \circ \beta} \qquad \mathrm{id}\,\mu \triangleright \mu; \mathrm{id}$$

$$\frac{\phi\mu \triangleright \mu_0; \beta \qquad \psi\mu' \triangleright \mu_0'; \beta'}{(\phi \star \psi)\mu \star \mu' \triangleright (\mu_0 \star \mu_0'); \beta \star \beta'} \qquad \mathrm{id}_\mathbf{1}\mathrm{id}_\mathbf{1} \triangleright \mathrm{id}_\mathbf{1}; \mathrm{id}_\mathbf{1}$$

$$\alpha(\mu \star (\mu' \star \mu'')) \triangleright (\mu \star \mu') \star \mu''; \alpha$$

$$\alpha^{-1}((\mu \star \mu') \star \mu'') \triangleright \mu \star (\mu' \star \mu''); \alpha^{-1}$$

$$\lambda\mu \;\triangleright\; \mathbf{1} \star \mu; \lambda \qquad \lambda^{-1}(\mathbf{1} \star \mu) \;\triangleright\; \mu; \lambda^{-1}$$

$$\gamma(\mu \star \nu) \;\triangleright\; (\nu \star \mu); \gamma$$

$$\xi\mu \;\triangleright\; \xi \circ \mu; \mathrm{id} \; (\xi \text{ constructor})$$

$$\frac{\phi\mu \;\triangleright\; \nu; \beta}{\pi_0(\langle\phi, \psi\rangle \circ \mu) \;\triangleright\; \nu; \beta} \qquad\qquad \frac{\psi\mu \;\triangleright\; \nu; \beta}{\pi_1(\langle\phi, \psi\rangle \circ \mu) \;\triangleright\; \nu; \beta}$$

$$\frac{\phi\mu \;\triangleright\; \nu; \beta}{[\phi, \psi](\kappa_0 \circ \mu) \;\triangleright\; \nu; \beta} \qquad\qquad \frac{\psi\mu \;\triangleright\; \nu; \beta}{[\phi, \psi](\kappa_1 \circ \mu) \;\triangleright\; \nu; \beta}$$

$$\frac{\phi(\mu \star \mu') \;\triangleright\; \nu; \beta}{\mathrm{ev}((\mathrm{cur}(\phi) \circ \mu) \star \mu') \;\triangleright\; \nu; \beta}$$

$\square$

REMARK. The definition of $\phi\mu \;\triangleright\; \nu; \beta$ is such that in an ILC the following diagram commutes:

$$
\begin{array}{ccc}
A & \xrightarrow{\;\phi\;} & B \\[2pt]
\Big\uparrow{\scriptstyle\mu} & \circledast & \Big\uparrow{\scriptstyle\nu} \\[2pt]
X & \xrightarrow{\;\beta\;} & Y
\end{array}
$$

Note, that if we start from a discrete graph, then β is, modulo a permutation, uniquely determined by X and Y alone; this follows from our coherence result for symmetric monoidal categories (Chapter 11). The idea is that (a) arrangement combinators are "trivial" from a computational point of view, and (b) non-primitive types and arrows between non-primitive types serve to compute with primitive arrows as data.

15.5. THEOREM. *For each* $\phi : A \longrightarrow B$, $\mu \in \mathrm{Can}(A)$, *there are unique* $\nu \in \mathrm{Can}(B)$, *and unique primitive* β *such that* $\phi\mu \;\triangleright\; \nu; \beta$.
PROOF Similar to the case of CCC's. We define for each A a set $[\![A]\!]$, the *computability type* of A. We write $\phi\mu \downarrow \mathcal{C}$ if $\mu \in \mathrm{Can}(A)$, and there are $\nu \in \mathcal{C}$ and a primitive β such that $\phi\mu \;\triangleright\; \nu; \beta$.

(i) $[\![A]\!] := \{\mathrm{id}_A\}$ for A atomic,

(ii) $[\![\mathbf{1}]\!] := \{\mathrm{id}_{\mathbf{1}}\}$,

(iii) $[\![A \star B]\!] := \{\mu \star \nu : \mu \in [\![A]\!], \; \nu \in [\![B]\!]\}$,

(iv) $[\![A \multimap B]\!] := \{(\text{cur}(\phi) \circ \mu : X \longrightarrow (A \multimap B)) :$
$\forall \nu \in [\![A]\!](\phi(\mu \star \nu) \downarrow [\![B]\!])\}$,

(v) $[\![A \sqcap B]\!] := \{\langle \phi, \psi \rangle \circ \mu \ : \ \phi : X \longrightarrow A, \ \psi : X \longrightarrow B,$
$\phi\mu \downarrow [\![A]\!], \ \psi\mu \downarrow [\![B]\!]\}$,

(vi) $[\![\top]\!] := \{\top_A \circ \mu\}$,

(vii) $[\![A \sqcup B]\!] := \{\kappa_0 \circ \mu : \mu \in [\![A]\!]\} \cup \{\kappa_1 \circ \mu : \mu \in [\![B]\!]\}$,

(viii) $[\![\bot]\!] := \emptyset$.

The remainder of the proof is contained in two lemmas.

LEMMA. *For each* $\phi : A \longrightarrow B$, $\mu \in [\![A]\!]$, *one has* $\phi\mu \downarrow [\![B]\!]$.
PROOF. By induction on the construction of ϕ. We check some cases.
Case 1. Let $\psi \circ \phi$ be given, with $\mu \in [\![A]\!]$. Then

$$\phi\mu \ \triangleright \ \nu; \beta \text{ for some } \nu \in [\![B]\!],$$
$$\psi\mu \ \triangleright \ \lambda; \beta' \text{ for some } \lambda \in [\![C]\!],$$

so $\psi\phi \ \triangleright \ \lambda; \beta' \circ \beta$.
Case 2. Let ev be given with $(\text{cur}(\phi) \circ \lambda) \star \mu$, where $\text{cur}(\phi) \circ \lambda \in [\![A \multimap B]\!]$, $\mu \in [\![A]\!]$, so if $\phi(\lambda \star \mu) \downarrow [\![B]\!]$, then $\text{ev}((\text{cur}(\phi) \circ \lambda) \star \mu) \triangleright \chi; \beta, \chi \in [\![B]\!]$ etc. $\square$

♠ EXERCISE. Do the other cases.

LEMMA. $[\![A]\!] = \text{Can}(A)$ *for all types* A.
PROOF. By induction on the definition of canonical combinator. Take e.g., $A \equiv B \multimap C$.

$$\text{Can}(B \multimap C) = \{\xi \circ \mu : X \longrightarrow B \multimap C\}$$

where $\xi : Y \longrightarrow (B \multimap C)$ is a constructor, say $\text{cur}(\phi)$, for $\phi : Y \star B \longrightarrow C$, and $\mu : X \longrightarrow Y \in \text{Can}(Y)$. $\mu \in [\![Y]\!]$ by IH; we have to show for $\nu \in [\![B]\!]$

$$\phi(\mu \star \nu) \downarrow [\![C]\!]$$

but this holds etc. $\square$

♠ EXERCISE. Complete the proof.

15.6. THEOREM. *The embedding* $\mathcal{C} \longrightarrow \text{ILC}(\mathcal{C})$ *is full.*
PROOF. Let A, B be atomic, and assume $\phi \in \text{Comb}(A, B)$. For any canonical $\mu \in \text{Can}(A)$ there is a unique $\nu \in \text{Can}(B)$ and a primitive β such that $\phi\mu \triangleright \nu; \beta$. Since $\mu = \text{id}_A$, $\nu = \text{id}_B$, or in a diagram:

$$
\begin{array}{ccc}
A & \xrightarrow{\;\phi\;} & B \\[2pt]
\Big\uparrow{\scriptstyle\mathrm{id}_A} & \circledast & \Big\uparrow{\scriptstyle\mathrm{id}_B} \\[2pt]
A & \xrightarrow{\;\beta\;} & B
\end{array}
$$

it follows that $\phi = \beta$, so ϕ is a primitive combinator from A to B; but all primitive combinators between atomic types are in fact in $\mathcal{C}$. $\square$

16 The categorical and linear machine

16.1. The categorical machine

This machine is based on the idea of calculating with categorical combinators of a CCC. At any stage in the computing process, the categorical machine has a *code* (a string of instructions to be executed), *active data* (the piece of information to be treated next) and a *stack* (a string of data in the memory). The basic instructions of the code are of the forms

$$\text{fst, snd, push, swap, cons, ev, cur}(C)$$

with C a string of code. The action of each of these basic instructions is indicated in the table below; we use $\frown$ for string concatenation.

Before execution			After execution		
code	active data	stack	code	active data	stack
fst $\frown C$	(u, v)	S	C	u	S
snd $\frown C$	(u, v)	S	C	v	S
push $\frown C$	u	S	C	u	$u \frown S$
swap $\frown C$	u	$v \frown S$	C	v	$u \frown S$
cons $\frown C$	u	$v \frown S$	C	(v, u)	S
ev $\frown C$	$(C' \cdot u, v)$	S	$C' \frown C$	(u, v)	S
cur$(C') \frown C$	u	S	C	$C' \cdot u$	S

16.2. Implementation of combinators

To see how the action of combinators can be implemented, we give some illustrations. Each combinator ϕ is translated into a string of code $[\![\phi]\!]$, the implementation of ϕ. We may think of the values appearing in the columns headed "active data" as values in the sense of 14.10. The implementation of a combinator, applied to a value in the second column, should as result yield a value as in Sections 14.11–14.14. Composition $\phi \circ \psi$ corresponds to concatenation of codes $[\![\psi]\!] \frown [\![\phi]\!]$ (order reversed!). Suppose we already know how to implement ϕ, ψ; we show how to implement $\langle \phi, \psi \rangle$. We write $[\![\phi]\!]u$ for the value resulting from execution of code $[\![\phi]\!]$ on active data u. $[\![\langle \phi, \psi \rangle]\!]u$ is computed by

- computing ϕ, preserving u;

- computing ψu;

- forming a pair $\langle \phi u, \psi u \rangle$.

Then $[\![\langle \phi, \psi \rangle]\!]$ becomes in code

$$\text{push} \frown [\![\phi]\!] \frown \text{swap} \frown [\![\psi]\!] \frown \text{cons}.$$

We show the action by exhibiting code, environment and stack in successive stages:

push $\frown [\![\phi]\!] \frown$ swap $\frown [\![\psi]\!] \frown$ cons $\frown C$	u	S
$[\![\phi]\!] \frown$ swap $\frown [\![\psi]\!] \frown$ cons $\frown C$	u	$u \frown S$
$\cdots$	$\cdots$	$\cdots$
swap $\frown [\![\psi]\!] \frown$ cons $\frown C$	$[\![\phi]\!]u$	$u \frown S$
$[\![\psi]\!] \frown$ cons $\frown C$	u	$([\![\phi]\!]u) \frown S$
$\cdots$	$\cdots$	$\cdots$
cons $\frown C$	$[\![\psi]\!]u$	$[\![\phi]\!]u \frown S$
C	$([\![\phi]\!]u, [\![\psi]\!]u)$	S

fst and snd implement the action of π_0 and π_1 respectively. $\text{cur}([\![\phi]\!]) \frown$ swap $\frown$ cons implements the action of the combinator $\text{cur}(\phi)$. The combinator ev is implemented by the instruction ev, i.e., $[\![\text{ev}]\!] = \text{ev}$.

$\text{ev}(([\![\phi]\!] \cdot u, v)$ is computed by forming the pair (u, v) and computing ϕ for this. In a table

ev $\frown C$	$(([\![\phi]\!] \cdot u), v)$	S
$[\![\phi]\!] \frown C$	(u, v)	S
$\cdots$	$\cdots$	$\cdots$
C	$[\![\phi]\!](u, v)$	S

16.3. REMARKS. (i) The implementation given uses lazy evaluation for values of function types, but eager evaluation for products. At the expense of more primitive instructions, lazy evaluation of products might also have been used for products. In particular, we may introduce closures $\mathrm{cur}(C') \cdot u$, $\langle C', C'' \rangle \cdot u$ with instructions as specified below.

Before execution			After execution		
code	active data	stack	code	active data	stack
$\mathrm{cur}(C') \frown C$	u	S	C	$\mathrm{cur}(C') \cdot u$	S
$\langle C', C'' \rangle \frown C$	u	S	C	$\langle C', C'' \rangle \cdot u$	S
$\mathrm{ev} \frown C$	$(\mathrm{cur}(C') \cdot u, v)$	S	$C' \frown S$	(u, v)	S
$\mathrm{fst} \frown C$	$\langle C', C'' \rangle \cdot u$	S	$C' \frown C$	u	S
$\mathrm{snd} \frown C$	$\langle C', C'' \rangle \cdot u$	S	$C'' \frown C$	u	S

For further comments see under the linear machine.

(ii) Lafont 1989 has an additional basic instruction ret ("return") and different instructions for ev and cur:

Before execution			After execution		
code	active data	stack	code	active data	stack
$\mathrm{cur}(C') \frown C$	u	S	C	$C' \cdot u$	S
$\mathrm{ev} \frown C$	$(C' \cdot u, v)$	S	C'	(u, v)	$\langle C \rangle \frown S$
$\mathrm{ret} \frown C'$	u	$\langle C \rangle \frown S$	$C \frown C'$	u	S

In this version the code steadily decreases during execution; on the other hand, code is temporarily dumped on the stack. In the instruction for "ev" we dump the code C as a singleton sequence $\langle C \rangle$, so that when we arrive at the return instruction, it is clear how much code has to be returned from the stack. We show the working of $\mathrm{cur}[\![\phi]\!] \frown \mathrm{swap} \frown \mathrm{cons} \frown \mathrm{ev} \frown C$ in this version:

$\mathrm{cur}[\![\phi]\!] \frown \mathrm{swap} \frown \mathrm{cons} \frown \mathrm{ev} \frown C$	u	$v \frown S$
$\mathrm{swap} \frown \mathrm{cons} \frown \mathrm{ev} \frown C$	$[\![\phi]\!]) \cdot u$	$v \frown S$
$\mathrm{cons} \frown \mathrm{ev} \frown C$	v	$([\![\phi]\!] \cdot u) \frown S$
$\mathrm{ev} \frown C$	$([\![\phi]\!] \cdot u, v)$	S
$[\![\phi]\!] \frown \mathrm{ret}$	(u, v)	$\langle C \rangle \frown S$
$\cdots$	$\cdots$	$\cdots$
ret	$[\![\phi]\!](u, v)$	$\langle C \rangle \frown S$
C	$[\![\phi]\!](u, v)$	S

The empty string can play the role of the return instruction; if we arrive at the empty code, it is time to fetch dumped code from the stack.

(iii) For simplicity we have left out the construct $[\phi, \psi]$ and the injections κ_0, κ_1, but it is not difficult to add them.

♠ EXERCISE. Add implementations for $[\,]$ and the injections κ_0, κ_1.

16.4. The linear abstract machine

This section is based on Lafont 1988a, 1988b. In a similar way we can describe a linear machine with basic instructions

> split, cons, xsplit, xcons,
> asl (associate left), asr (associate right),
> insl (insert left), dell (delete left),
> exch, ξ for any constructor ξ,
> ev, fst, snd.

Constructors are pieces of code of the form $\mathrm{cur}(C)$ or $\langle C, C' \rangle$ with C, C' code. The effect of the basic instructions is given in the table below. The basic constructions are not all independent; in particular, exch is definable as split $\frown$ xcons, xcons as exch $\frown$ cons, xsplit as exch $\frown$ split. Dropping one of these in favor of the others is arbitrary however.

	Before execution			After execution	
code	active data	stack	code	active data	stack
split $\frown$ C	(u,v)	S	C	u	$v \frown S$
cons $\frown$ C	u	$v \frown S$	C	(u,v)	S
xsplit $\frown$ C	(u,v)	S	C	v	$u \frown S$
xcons $\frown$ C	u	$v \frown S$	C	(v,u)	S
asl $\frown$ C	$(u,(v,w))$	S	C	$((u,v),w)$	S
asr $\frown$ C	$((u,v),w)$	S	C	$(u,(v,w))$	S
insl $\frown$ C	u	S	C	$((\,),u)$	S
dell $\frown$ C	$((\,),u)$	S	C	u	S
exch $\frown$ C	(u,v)	S	C	(v,u)	S
$\xi \frown$ C	u	S	C	$\xi \cdot u$	S
ev $\frown$ C	$(\mathrm{cur}(C') \cdot u, v)$	S	$C' \frown C$	(u,v)	S
fst $\frown$ C	$\langle C', C'' \rangle \cdot u$	S	$C' \frown C$	u	S
snd $\frown$ C	$\langle C', C'' \rangle \cdot u$	S	$C'' \frown C$	u	S

To see the connection with the evaluation of combinators in the preceding chapter, for the case of a discrete category $\mathcal{C}$, where identities are the only atomic combinators, think of the values appearing in the columns "active data" as ranging over the canonical combinators of the preceding chapter. If u, v are canonical, so is $u \star v$, represented by (u,v); the canonical combinator id_1 is represented by ( ). If C is the code for a combinator $\phi : B \longrightarrow A$, and $u : X \longrightarrow B$ is canonical, then $\mathrm{cur}(C') \cdot u$ represents the canonical combinator $\phi \circ u$. If C, C' are codes for $\phi : B \longrightarrow A$ and $\phi' : B \longrightarrow A'$ respectively, and $u : X \longrightarrow B$ is canonical, then $\langle C, C' \rangle \cdot u$ represents the canonical combinator $\langle \phi, \phi' \rangle \cdot u$.

For simplicity we consider linear logic $\mathbf{ILL}_0$ without $\sqcup$. The implementation of the various combinators can now be given in the same way as for the categorical machine.

Parallel composition of combinators is translated sequentially, i.e., we code $\phi \star \psi$ as $(\mathrm{id} \star \phi) \circ (\psi \star \mathrm{id})$, that is to say as

$$\text{split} \frown [\![\psi]\!] \frown \text{xcons} \frown \text{split} \frown [\![\phi]\!] \frown \text{xcons};$$

$[\![\mathrm{id}]\!]$ is the empty string of code.

The implementation of $\langle \phi, \psi \rangle$ is in this case simply $[\![\langle \phi, \psi \rangle]\!] :=$ $\langle [\![\phi]\!], [\![\psi]\!] \rangle$. We illustrate the effect of "fst":

$\langle [\![\phi]\!], [\![\psi]\!] \rangle \frown \text{fst} \frown C$	u	S
$\text{fst} \frown C$	$\langle [\![\phi]\!], [\![\psi]\!] \rangle \cdot u$	S
$[\![\phi]\!] \frown C$	u	S
$\cdots$	$\cdots$	$\cdots$
C	$[\![\phi]\!]u$	S

16.5. REMARKS. (i) Note that none of the instructions for the linear machine permits the duplication of values, in contrast to the case of the categorical machine, where the instruction "push" copies the value of the active data to the stack. Canonical values of primitive type are not thrown away, though sometimes pieces of code are.

(ii) The case of the non-discrete graph $\mathcal{C}$ can be included by having an $[\![\delta]\!]$ instruction for each atomic combinator $\delta : A \longrightarrow B$:

Before execution			After execution		
code	active data	stack	code	active data	stack
$[\![\delta]\!] \frown C$	id_A	S	C	id_B	S

(iii) The comments on alternative forms of the instructions using "ret" or the empty code also apply here.

♠ EXERCISE. Add implementations for $[\,]$ and the injections κ_0, κ_1 also in this case.

17 Proofnets for the multiplicative fragment

17.1. In this chapter we present Girard's notion of proofnet for the multiplicative fragment; references are Girard 1987, 1991, Danos and Regnier 1989, Roorda 1990,1991, Gallier 1991. We start from the one-sided version of the sequent calculus with axioms and rules

$$\mathbf{1} \qquad \frac{\Gamma}{0,\Gamma} \qquad A,\sim A$$

$$\star \frac{A,\Gamma \quad B,\Delta}{A \star B,\Gamma,\Delta} \qquad + \frac{A,B,\Gamma}{A+B,\Gamma} \qquad \text{Cut} \frac{\Gamma,A \quad \Delta,\sim A}{\Gamma,\Delta}$$

Cutfree sequent proofs may differ in the order of the application of the rules, e.g.,

$$\frac{\dfrac{A,\sim A \quad B,\sim B}{A \star B,\sim A,\sim B} \quad C,\sim C}{\dfrac{(A \star B) \star C,\sim A,\sim B,\sim C}{(A \star B) \star C,\sim A + \sim B,\sim C}} \qquad \frac{\dfrac{\dfrac{A,\sim A \quad B,\sim B}{A \star B,\sim A,\sim B}}{A \star B,\sim A + \sim B} \quad C,\sim C}{(A \star B) \star C,\sim A + \sim B,\sim C}$$

represent "essentially" the same proof: only the order of the application of the rules differs. The proofs also exhibit a lot of redundancy inasmuch the inactive formulas are copied many times.

144

Proofnets were introduced in order to remove such redundancies and to find a unique representative for equivalent sequent calculus proofs.

17.2.   A proof structure (in the context-free fragment) is a graph with nodes labeled by formulas or the cut symbol "cut", built from the following components:

- isolated nodes **1**, **0**;

- axiom links $A \text{———} \sim A$;

- cut links $A \text{——} \text{cut} \text{——} \sim A$;

- logical rules:
  $\star$-link $A \xrightarrow{\;\star\;} A \star B \xrightarrow{\;\star\;} B$ and
  +-link $A \xrightarrow{\;+\;} A + B \xrightarrow{\;+\;} B$.

So edges with cut, $\star$, + always appear in pairs. We do not actually need to label the nodes; the combinatorial behaviour will be equally well determined by labeling the edges with 0 (for axiom links), 1 (for +-links) and 2 (for $\star$- and cut-links). More precisely we define proof structures as follows:

DEFINITION.   *Proof structures* ν and the set of *terminal nodes* $\mathrm{TN}(\nu)$ of ν are defined simultaneously. Let ν, A, B, $\ldots D$ indicate a proof structure (PS) ν with some of its terminal nodes labeled A, B, $\ldots D$. We shall indulge in a slight abuse of notation in frequently using the labels to designate the nodes. Proof structures are generated by the clauses:

(i) single nodes **0**, **1** are PS, with terminal nodes **0**, **1** respectively;

(ii) if ν, μ are PS, then so is $\nu \cup \mu$, with $\mathrm{TN}(\nu \cup \mu) = \mathrm{TN}(\nu) \cup \mathrm{TN}(\mu)$;

(iii) $A \text{———} \sim A$ is a PS (axiom), with terminal nodes $A, \sim A$;

(iv) if $\nu, A, \sim A$ is a PS, so is the graph obtained by adding edges and the symbol "cut" $A \text{——} \text{cut} \text{——} \sim A$; the new terminal nodes are $\mathrm{TN}(\nu) \cup \{\text{cut}\} \setminus \{A, \sim A\}$ (i.e., terminal nodes of ν except $A, \sim A$, and "cut" added);

(v) if ν, A, B is a PS, then so are the graphs obtained by adding two edges and a node $A \text{———} A \star B \text{———} B$ or $A \text{———} A + B \text{———} B$ ($\star$-link and +-link). The terminal nodes are $\mathrm{TN}(\nu)$ with A, B omitted and $A \star B$, resp. $A + B$ added.

A notation for PS which is closer to deduction notation is obtained by the following version of the definition (with the obvious clauses for terminal nodes):

(a) $\overline{A \qquad \sim\!A}$ is a PS (axiom link);

(b) $\overline{1}, \overline{0}$ are PS;

(c) the union of two PS is a PS;

(d) connecting terminal nodes A, B in a PS by

$$\frac{A \quad B}{A \star B} \text{ or } \frac{A \quad B}{A + B}$$

gives a new PS (adding a $\star$-link and a $+$-link respectively);

(e) connecting terminal nodes A, $\sim\!A$ in a PS by

$$\frac{A \quad \sim\!A}{\text{cut}}$$

gives a new PS. $\square$

EXAMPLE. The two sequent proofs at the beginning of this chapter are both represented by the following PS.

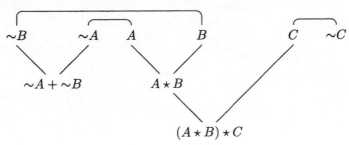

Another graphic representation of the same PS is as follows:

$$\frac{\overset{\frown}{\sim\!A \quad \sim\!B} \quad A \quad B}{\dfrac{\sim\!A + \sim\!B \qquad \dfrac{A \star B}{(A \star B) \star C} \quad C \quad \sim\!C}{}}$$

A certain subset of the PS corresponds in an obvious way to sequent deductions.

17.3. Definition.   *Inductive PS* (IPS) are obtained by the following clauses:

(i) **1** is an IPS, TN(ν) = **1**;

(ii) if ν is an IPS, then so is ν, **0** (node **0** added, no new edges); TN(ν, **0**) = $\nu \cup \{$**0**$\}$

(iii) $A \text{———} {\sim}A$ is an IPS (axiom), with $A, {\sim}A$ as terminal nodes;

(iv) if ν, A and $\nu', {\sim}A$ are IPS, then so is $\nu, A \text{——} \text{cut} \text{——} {\sim}A, \nu'$ (cut link: two new edges and a node labeled "cut"), terminal nodes are (TN(ν) $\cup$ TN(ν') $\cup \{$cut$\}$) $\setminus \{A, {\sim}A\}$;

(v) if ν, A and ν', B are IPS, then so is $\nu, A \text{——} A \star B \text{——} B, \nu'$ (two new edges and a node $A \star B$), terminal nodes are (TN(ν) $\cup$ TN(ν') $\cup \{A \star B\}$) $\setminus \{A, B\}$;

(vi) if ν, A, B is an IPS, then so is

$$\nu, A, B$$
$$\setminus \, /$$
$$A + B$$

(two new edges and a node $A + B$), terminal nodes as in the corresponding clause of the preceding definition. $\square$

N.B. The example above of a PS is in fact an IPS.

We now address the question, whether there is an intrinsic criterion to decide whether a PS is in fact an IPS, i.e., can be generated from a sequent calculus deduction. For this the notion of a "trip" has been devised. A trip is a route along the nodes of a PS, such that each formula occurrence A may be passed in two directions (written as $A \downarrow$, $A \uparrow$), according to certain instructions for passing the links. From now on, we strictly restrict attention to the fragment with $\star$, $+$ and Cut rule; we drop **0**, **1**.

17.4. Definition.   (Travel instructions for trips)
(i) Axiom link: $A \uparrow$ followed by ${\sim}A \downarrow$, $A \downarrow$ followed by ${\sim}A \uparrow$.

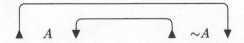

(ii) Terminal node: $A \downarrow$ followed by $A \uparrow$.

(iii) $\star$-link $A \star B$.
Switch on L (left): $B \downarrow$ followed by $A \uparrow$, $A \downarrow$ followed by $A \star B \downarrow$, $A \star B \uparrow$ followed by $B \uparrow$;

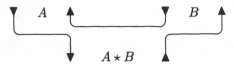

Switch on R (right) :
$A \downarrow$ followed by $B \uparrow$, $B \downarrow$ followed by $A\star$; $B \downarrow$, $A \star B \uparrow$ followed by $A \uparrow$

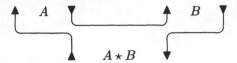

(iv) $+$-link $A + B$.
Switch on L (left): $A \downarrow$ followed by $A + B \downarrow$, $A + B \uparrow$ followed by $A \uparrow$, $B \downarrow$ followed by $B \uparrow$;

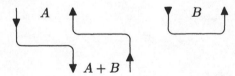

Switch on R (right):
$B \downarrow$ followed by $A + B \downarrow$, $A + B \uparrow$ followed by $B \uparrow$, $A \downarrow$ followed by $A \uparrow$.

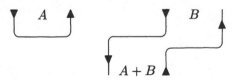

(v) Cut-link: as $\star$-link, where "Cut" is treated as a terminal formula.
□

17.5. DEFINITION. A *trip* is a sequence $A_1 I_1$, $A_2 I_2$, $\ldots A_n I_n$ according to the travel instructions, where the A_i are formulas, and each I_i is either ↑ or ↓, and which cannot be extended without repetition. A trip in a connected PS with n nodes is a *longtrip* if it has length $2n$. A longtrip in a PS with connected parts ν, ν', ν'', ... is a set of longtrips for ν, ν', ν'', ... respectively. A *proofnet* is a PS in which all trips are longtrips. □

17.6. THEOREM. *A connected PS is an IPS iff every trip is a longtrip.*

♠ EXERCISE. Prove that an IPS satisfies the longtrip condition.

It is easy to verify that each IPS is a PS satisfying the longtrip condition. For the converse we have to work much harder. In order to prove this converse, we first replace the longtrip condition by another, equivalent, condition and prove a lemma.

17.7. DEFINITION. A *switching* of a +-link is the replacement of a +-link

$$A \longrightarrow A + B \longrightarrow B$$

either by the configuration

$$A \longrightarrow A + B \qquad B$$

or by the configuration

$$A \qquad A + B \longrightarrow B.$$

A *switching of a PS* is the graph resulting from the structure by taking a switching for each +-link. □

We use the term "switching" since the choice of a switching of a +-link as defined here corresponds to a switching for the link in the sense of the travel instructions:

$$A \longrightarrow A + B \qquad B$$

corresponds to switch "L", and

$$A \qquad A + B \longrightarrow B$$

to switch "R". A switching of a proof structure corresponds to a setting of all switches at +-links.

17.8. LEMMA. *The longtrip condition is equivalent to the condition that each switching of the proof structure is an acyclic and connected graph (in other words, a tree).*

PROOF. Here and in the sequel we shall disregard in the proofs the cut links, since these can be treated exactly as $\star$-links. We may picture proof structures and switchings as graphs with a top layer of formulas connected by axiom links; in the top layer links may cross. Below the top layer the structure is tree-like, without crossings. So a cycle in a switching is typically something like:

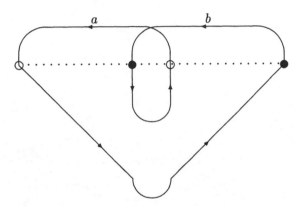

where the parts a, b are axiom links, and the dotted line separates the top layer from the bottom layer. Below the top layer a cycle first goes steadily down, then up until the top layer has been reached. So each dip into the bottom part has a (local) minimum corresponding to a unique $\star$-link. Clearly it is possible to set the switches of the $\star$-links so that one obtains a shorttrip following the cycle; for a $\star$-link at a local minimum, the trip passes through the premises but not through the conclusion.

Conversely, it is easy to see that for an acyclic and connected graph only longtrips can arise. □

♠ EXERCISE. Prove that an IPS is a proofnet using the tree criterion.

We now turn to the proof that every proofnet is an IPS (i.e., can be obtained from a sequent proof). The argument below closely follows Girard 1991, restricted to proofnets without quantifiers; it is an adaptation of the original proof formulated on the basis of the longtrip condition.

17.9. LEMMA. *Let ν be a proofnet, with a terminal formula $A + B$, part of an $+$-link. Deleting the formula with its connecting edges results in another proofnet.*

PROOF. Obvious; acyclicity cannot be spoiled by removing links, and connectedness of any switching of ν cannot depend on the presence of $A + B$, by the definition of switching. $\square$

17.10. DEFINITION. A terminal formula of a connected PS is called *splitting* if removal of the formula with its connecting edges leads to two disjoint structures.

A maximal connected subset of a graph is called a *component*. For the complement of a set of nodes X within a given graph we write X^c. $\square$

Our aim is now to show that a proofnet with all terminal formulas conclusion of a $\star$-link has a splitting terminal formula. This is done in a series of lemmas involving the notion of empire, to be defined below.

17.11. DEFINITION. (Empire) Let ν be a proofnet, A a formula occurrence in ν, S a switching of ν with graph $S(\nu)$, and let $S(\nu, A)$ be the subgraph of $S(\nu)$ obtained as follows: if A is a premise of a link with conclusion A', and the edge $A \text{———} A'$ is in $S(\nu)$, S' is $S(\nu)$ with the edge $A \text{———} A'$ deleted; otherwise $S' = S(\nu)$. S' has at most two components; let $S(\nu, A)$ be the component containing A. The *empire* $e(A)$ of A is the intersection of the $S(\nu, A)$ for all possible switchings S of ν. $\square$

17.12. LEMMA. *Let ν be a proofnet. Then*

(i) *If $B \text{———} \sim B$ is an axiom link in ν, then $B \in e(A)$ iff $\sim B \in e(A)$.*

(ii) *If $B \text{———} B \star C \text{———} C$ is a $\star$-link in ν, and $B, C \not\equiv A$, then $B \in e(A)$ iff $B \star C \in e(A)$ iff $C \in e(A)$.*

(iii) *If $B \text{———} B + C \text{———} C$ is a $+$-link in ν, and $B, C \not\equiv A$, then $B + C \in e(A)$ iff ($B \in e(A)$ and $C \in e(A)$).*

(iv) *If A is premise of a $+$-link or $\star$-link, then the conclusion of the link does not belong to $e(A)$.*

PROOF. (i),(ii),(iv) are immediate from the definition. As to (ii), if $B, C \not\equiv A$, then $B \in e(A)$, $C \in e(A) \Rightarrow B + C \in e(A)$. For the converse we argue by contradiction: let $B + C \in e(A)$ be the conclusion of a $+$-link, and assume $C \notin S(\nu, A)$ for some switching S. A must be premise of a link with conclusion A', $A \text{———} A'$ is selected in S (otherwise $S(\nu, A)$ would have a single component), and C is in the same component as A', hence $S(\nu, A)$ contains the edge $B \text{———} B + C$.

Consider now the switching S' which differs from S only in that it contains the edge $C \relbar\joinrel\relbar B + C$ instead of $B \relbar\joinrel\relbar B + C$. Then still $B+C \in S'(\nu, A)$ since $B+C \in e(A)$, and now $B+C$ is via C connected with A' in $S'(\nu, A)$, not via $A \relbar\joinrel\relbar A'$, (since the connection between C and A' in $S(\nu, A)$ persists in $S'(\nu, A)$); but then S' contains a cycle. □

17.13. LEMMA. *For any proofnet ν and formula occurrence A in ν there is a suitable switching S such that $S(\nu, A) = e(A)$.*

PROOF. Let us call a switching as postulated in the lemma a *principal* switching for A. We obtain a principal switching as follows: if A is the premise of a +-link with conclusion A', include the edge $A \relbar\joinrel\relbar A'$ in the switching. For a +-link $B \relbar\joinrel\relbar B + C \relbar\joinrel\relbar C$, $B, C \not\equiv A$, with $B + C \notin e(A)$, by the preceding lemma at least one of the premises is not in $e(A)$; include in S the edge connecting $B + C$ with such a premise.

In this way the only edge (if any) in S connecting $e(A)$ with its complement is the edge $A \relbar\joinrel\relbar A'$ (if A is premise of a +-link); hence $e(A) = S(\nu, A)$. □
Observe that this construction by no means uniquely determines S.

17.14. LEMMA. *(Nesting lemma) Let A, B be distinct formula occurrences in a proofnet ν, and assume $B \notin e(A)$. Then*

(i) $A \in e(B) \Rightarrow e(A) \subset e(B)$;

(ii) $A \notin e(B) \Rightarrow e(A) \cap e(B) = \emptyset$.

PROOF. Assume $B \notin e(A)$. We specialize the construction of a principal switching S for B as follows.

For a +-link with conclusion $C \equiv C_1 + C_2$ ($C_1, C_2 \not\equiv A$) but not in $e(A)$, include in S an edge between C and a premise not in $e(A)$.

If A is premise of a link with conclusion $A' \in e(B)$, include $A \relbar\joinrel\relbar A'$ in S.

We note that $S(\nu)$ does not contain edges connecting $e(A) \cap e(B)$ and $e(A)^c \cap e(B)$ except possibly an edge $A \relbar\joinrel\relbar A'$. Since S is principal for B, $S(\nu, B) = e(B)$. Two possibilities arise.

If $A \in e(B)$, then, since $B \notin e(A)$, $B \in e(B)$, there is an edge between $e(A)$ and $e(A)^c$ within $e(B)$, and this can only be the edge $A \relbar\joinrel\relbar A'$; from this $S(\nu, A) \subset S(\nu, B)$ and since $e(A) \subset S(\nu, A)$, we see that $e(A) \subset e(B)$.

If $A \notin e(B)$, an edge between $e(A)$ and $e(A)^c$ within $e(B)$ is excluded, and since $B \notin e(A)$, no formula of $e(A)$ can belong to $e(B)$. □

17.15. LEMMA. *(Splitting lemma) A proofnet with all terminal formulas part of a $\star$-link has a splitting formula.*

PROOF. Let $A_i \text{———} A_i \star B_i \text{———} B_i$ $(1 \leq i \leq n)$ be the terminal $\star$-links. Among the empires $e(A_i)$, $e(B_j)$ at least one must be maximal w.r.t. inclusion, say $e(A_i)$. We define the *border* of $e(A_i)$ as consisting of (the occurrence) A_i and the formula occurrences in $e(A_i)$ which are either terminal or premise of a link with conclusion not in $e(A_i)$.

We claim that in fact the border of $e(A_i)$ contains only A_i and terminal formulas of ν. For if C is any other formula in the border, it is premise of a $+$-link with conclusion $C' \notin e(A_i)$. However, below C and C' there is a terminal $\star$-link, and C, C' are above one of the premises of such a link, say A_j (we use "below" and "above" as in 17.8: alternatively, one could say that C, C' are hereditary premises of A_j).

Then $A_j \notin e(A_i)$, otherwise $C' \in e(A_i)$ (as is easy to verify from Lemma 17.12); and since also $C \in e(A_j)$, it follows that $e(A_i) \cap e(A_j) \neq \emptyset$. But then by the preceding lemma $e(A_i) \subset e(A_j)$, and this contradicts the maximality of $e(A_i)$.

It follows from the claim just established that, for all switchings S, $e(A_i) = S(\nu, A_i)$. From this in turn it follows that necessarily $e(B_i) = S(\nu, B_i)$ for all switchings S.

Now if $A \text{———} A \star B \text{———} B$ is any $\star$-link, any edge between $e(A)$ and $e(B)$ would immediately lead to a cycle in a suitable switching $S'(\nu)$ by creating a connection in $S'(\nu)$ between A and B not via $A \star B$. It follows that $e(A_i)$ and $e(B_i)$ are *only* connected via $A_i \star B_i$, and that $e(A_i)$ and $e(B_i)$ are two components filling up all of ν minus $A_i \star B_i$ with its adjoining edges. $\square$

17.16. THEOREM. *Any proofnet is an IPS.*

PROOF. By induction on the number of nodes in a proofnet ν. If ν has a $+$-link with terminal conclusion A, apply Lemma 17.9. ν with A and adjoining edges deleted is a proofnet ν'; by the induction hypothesis ν' is an IPS, and therefore ν is.

If ν has no terminal $+$-link, we apply the splitting lemma to find a terminal $A \star B$, part of a $\star$-link, which is splitting; deleting $A \star B$ with adjoining edges results in two smaller disjoint proofnets to which the induction hypothesis applies; hence ν is an IPS. $\square$

17.17. The 2-property for BCI-logic

The following argument can also be given for the sequent calculus, but is particularly direct using proofnets. Since classical linear logic is conservative over the intuitionistic $\multimap$-fragment of **ILL**, we can represent proofs of BCI-theorems by proofnets, where $A \multimap B$ is simply treated as $\sim A + B$. For any cutfree proofnet ν with only atomic instances of the axiom it is clear that ν remains a proofnet if we replace the proposition variables by distinct variables for each axiom. Thus we obtain:

PROPOSITION. *Any provable formula of BCI-logic can be obtained by substitution from a provable formula in which each propositional variable occurs exactly twice.*

18 The algorithm of cut elimination for proof nets

The present section is based on Girard 1988.

18.1. Representing a proofnet as a permutation

The aim of this chapter is to give an algorithmic interpretation of cut elimination for proofnets by means of permutations; permutations may be represented by matrices with a single 1 in each row and column, i.e., very special linear transformations. Thus the interpretation gives us a slight foretaste of Girard's "geometry of interaction", i.e., the interpretation of cut elimination in linear logic in terms of C^*-algebras (Girard 1989).

We strictly limit attention to proofnets for the multiplicative fragment with $\star$, $+$, cut. Assume ν to be a cutfree proofnet with a single conclusion A, and only *atomic* axiom links. Then ν may be regarded as consisting of the construction tree T_A of A (exhibiting the construction of A from its subformulas), where at the top nodes appear atoms P, $\sim P$ (P prime), together with a set of axiom links connecting the top nodes pairwise.

Clearly, T_A is the same for all such proofnets for A; the only differences occur in the axiom links. Suppose $b_1, \ldots, b_n$ is the set of top nodes. Let S be any switching of the links in T_A. Then S determines ·

155

a permutation f_S of $\{1, \ldots, n\}$ by:

> In the trip determined by S,
> the next top node after $b_i \downarrow$ is $b_{f_S(i)} \uparrow$.

Let us write $\mathrm{Per}(A)$ for the set of the f_S. There is also a permutation g_ν depending on ν:

$$g_\nu(i) = j \text{ if } b_i, b_j \text{ are linked by an axiom link.}$$

So g_ν has no fixed points and is involutory, i.e., $g_\nu(g_\nu(i)) = i$. We note

ORT $\qquad\qquad\qquad g_\nu \perp f_S$ for all $f_S \in \mathrm{Per}(A)$;

here $\sigma \perp \tau$ means that $\sigma\tau$ is cyclic of order n. This is a consequence of the longtrip condition.

18.2. LEMMA. *ORT is equivalent to the longtrip condition.*
PROOF. Assume that for some trip and some switching S, some compound formula A is visited downwards: $A \downarrow$, but not upwards. $A \downarrow$ comes from a top node $b_i = P \downarrow$ (or $\sim P \downarrow$). On the other hand, *another* trip starting at $A \uparrow$ say, with the same switching S, ends in $b_j = Q \uparrow$ (or $\sim Q \uparrow$). This means that applying $g_\nu f_S$, beginning with b_i, can never bring us to b_j, so $g_\nu f_S$ is not cyclic of order n. $\square$

18.3. NOTATION. For a set of permutations X, we write $X^\perp$ for the set of permutations σ such that $\sigma \perp \tau$ for all $\tau \in X$. $\square$
    We generalize the notion of a normal proof in what follows to the notion of a normal *preproof*, which is nothing but a permutation σ of $\{1, \ldots, n\}$ such that $\sigma \perp f_S$ for all f_S. Note that $\perp$ is symmetric, and that $X \subset Y \Rightarrow Y^\perp \subset X^\perp$.
    The notion of trip may be extended to preproofs in the obvious way, and the proof of the preceding chapter showing the equivalence between the longtrip condition and the condition that every switching is a tree, i.e., acyclic and connected, remains valid.

18.4. LEMMA. $\mathrm{Per}(\sim A)^\perp = \mathrm{Per}(A)^{\perp\perp}$.
PROOF. We first show $\mathrm{Per}(A)^{\perp\perp} \subset \mathrm{Per}(\sim A)^\perp$ by showing

(a) $\qquad\qquad\qquad \mathrm{Per}(A) \perp \mathrm{Per}(\sim A)$

Consider an axiom link $\overline{A \quad \sim A}$ and expand this into into a deduction of $A, \sim A$ with atomic axiom links in a canonical way, e.g for $A \equiv (B + C) \star D$

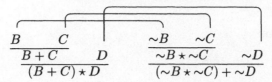

$$\frac{B \quad C}{\dfrac{B + C}{(B+C) \star D} \quad D} \qquad \frac{\sim B \quad \sim C}{\dfrac{\sim B \star \sim C}{(\sim B \star \sim C) + \sim D} \quad \sim D}$$

Let g be the involution interchanging corresponding atoms of A and $\sim A$. g describes a normal proof of $A + \sim A$. Label the atoms of A: $b_1, b_2, \ldots, b_n$ and of $\sim A$: $b_{n+1}, b_{n+2}, \ldots, b_{2n}$, such that $g(i) = i + n$. Any element $f \in \mathrm{Per}(A)$ is a permutation of $\{1, 2, \ldots, n\}$, and $f' \in \mathrm{Per}(\sim A)$ is a permutation of $\{n + 1, n + 2, \ldots, 2n\}$. Let $f \cup f'$ be the union defined on $\{1, 2, \ldots, 2n\}$; then $f \cup f' \perp g$, since g corresponds to a normal proof. But then $f \perp f''$, where $f''(i) := f'(i + n) - n$ is a permutation of $\{1, 2, \ldots, n\}$. To see this, note that

$$g(f \cup f')(i) = \begin{cases} f(i) + n \text{ for } i \leq n, \\ f'(i) - n \text{ for } i > n. \end{cases}$$

So, keeping in mind that $(g(f \cup f'))^2$ is cyclic of order n,

$$(g(f \cup f'))^2(i) = f'(f(i) + n) - n = f'' f(i).$$

Clearly, it must be the case that $f'' \perp f$. Next we prove $\mathrm{Per}(\sim A)^\perp \subset \mathrm{Per}(A)^{\perp\perp}$ by showing

(b) $$\mathrm{Per}(\sim A)^\perp \perp \mathrm{Per}(A)^\perp,$$

or if $g' \in \mathrm{Per}(\sim A)^\perp$, $g \in \mathrm{Per}(A)^\perp$, then $g \perp g'$. Consider now the following structure.

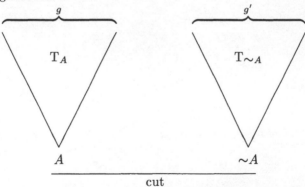

This structure has the longtrip property. This is easily verified by checking that every switching of the new structure is a tree.

Now we show that a cut-elimination step leads again to a structure satisfying the longtrip condition. Suppose $A \equiv B \star C$ and replace

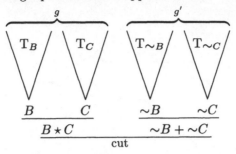

by

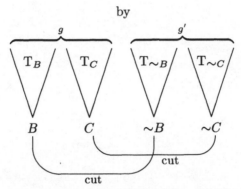

Again it is easy to see that a switching of the right hand structure is a tree, if switchings of the left hand structure are always trees.

Continuing the cut elimination, this leads to a structure with a chain of cuts

$$\cdots \frac{P_1 \quad \sim P_1}{\text{cut}} \quad \frac{P_2 \quad \sim P_2}{\text{cut}} \quad \frac{P_3 \quad \sim P_3}{\text{cut}} \cdots$$

such that $(g \cup g')$ maps $\sim P_n$ to P_{n+1}. Clearly, if P_i is in the domain of g, then $\sim P_i$ is in the domain of g' and vice versa. Hence gg' is cyclic of order n, so $g \perp g'$. $\square$

18.5. THEOREM. *Let ν be a proofnet with terminal formula A and cutformulas $B_1, B_2, \ldots, B_m$, and axiom-link permutation g_ν. ν can be transformed into a cutfree proofnet ν'' of*

$$C \equiv A + (B_1 \star \sim B_1) + \cdots + (B_m \star \sim B_m)$$

(replace cutlinks by $\star$-links, and combine the terminal formulas by $+$-links). $g_{\nu''} = g_\nu$.

Let the top nodes of T_C be numbered b_1, b_2, ..., b_n such that b_1, ... b_k derive from A and b_{k+1}, ... b_n from the $B_i, \sim B_i$, such that b_{2l} of B_i corresponds to b_{2l+1} of $\sim B_i$. Let h be the permutation:

$$h(j) = j \text{ for } j \leq k;$$
$$h(j) = j + 1 \text{ for } j > k, j - k \text{ odd};$$
$$h(j) = j - 1 \text{ for } j > k, j - k \text{ even}.$$

Define $g_{\nu'}$ for $i \leq k$ by

$$g_{\nu'}(i) = (g_\nu h)^{d_i}(i)$$

where d_i is the least positive number c such that

$$(g_\nu h)^c(i) \leq k.$$

Then $g_{\nu'}$ represents a cutfree proof of A.

PROOF. Note that, since $g_\nu h$ is a permutation, there is a $c > 0$ such that $(g_\nu h)^c(i) = i$, and hence $g_{\nu'}$ is well-defined. Let $f \in \mathrm{Per}(A)$; we have to prove $g_{\nu'} \perp f$. By the lemma f is a preproof of $\sim A$. Therefore we can obtain a preproof

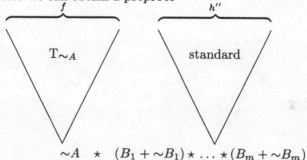

$$\sim A \quad \star \quad (B_1 + \sim B_1) \star \ldots \star (B_m + \sim B_m)$$

where the standard part has axiom links corresponding to the permutation h. Let h' coincide with h'' on the axioms of the standard part, and with f on arguments $\leq k$. By the lemma $h' \perp g_\nu$, so $g_\nu h'$ is cyclic. Now

$$(g_{\nu'} f)(i) = g_{\nu'}(f(i)) = (g_\nu h)^{d_{f(i)}} f(i),$$

so

$$g_\nu h f(i) > k, \ (g_\nu h)^2(f(i)) > k, \ldots (g_\nu h)^{d_{f(i)}-1} f(i) > k,$$

and hence, noting that $hf(i) = h'(i)$ for $i < k$,

$$\begin{aligned}
(g_{\nu'} f)(i) &= (g_\nu h')^{d_{f(i)}-1}(g_\nu h(f(i))) \\
&= (g_\nu h')^{d_{f(i)}-1}(g_\nu h'(i)) \\
&= (g_\nu h')^{d_{f(i)}}(i).
\end{aligned}$$

Then, if $u = (g_\nu h')^{d_{f(1)}}(1)$,

$$
\begin{aligned}
(g_{\nu'} f)^2(1) &= (g_\nu h)^{d_{f(u)}}(f(u)) \\
&= (g_\nu h')^{d_{f(u)}}(u) \\
&= (g_\nu h')^{d_{f(1)}+d_{f(u)}}(1).
\end{aligned}
$$

From this it follows that $g_{\nu'} f$ is cyclic since $g_\nu h'$ is cyclic.

Furthermore, if g_ν is not just a preproof but a proof, i.e., an involutory permutation generated by axiom links, then so is $g_{\nu'}$. For if we apply cut elimination to a proofnet for A with cuts, the final step in the elimination procedure shows a number of chains of atomic cuts:

$$
\frac{\overline{P_0 \quad \sim P_0}}{\vdots} \underset{\text{cut}}{\overline{P_1 \quad \sim P_1}} \underset{\text{cut}}{\overline{P_2 \quad \sim P_2}} \ldots \underset{\text{cut}}{\overline{P_n \quad \sim P_n}} \vdots
$$

All P_i are identical, and two occurrences connected by a cut link derive from a pair $B_i + \sim B_i$; P_0 and $\sim P_n$ derive from A. The transition from g_ν to $g_{\nu'}$ corresponds to the contraction of such sequences to axiom links connecting P_0 directly with $\sim P_n$. $\square$

18.6. Proof expressions for proofnets

Abramsky (1990) replaces proofnets by a kind of generalized term calculus, a calculus of *proof expressions*. The idea is to combine a term notation for proofs in the one-sided sequent calculus with a list of cuts occurring in the proof. Thus a proof expression is of the form $\Theta; \vec{t} : \Gamma$, where a $\vec{t} : \Gamma$ is a set, and Θ is a multiset of cuts, each cut written as $t \perp t'$ ($\perp$ is symmetric). A cutfree proof corresponds to an expression of the form $; \vec{t} : \Gamma$ (or $\Lambda; \vec{t} : \Gamma$ where as before Λ is the empty multiset). The assignment of proof expressions in the multiplicative$\{+, \star\}$-fragment is given by (x, y, z variables)

$$
\text{Ax} \quad ; x : A, x : \sim A \qquad \text{Cut} \; \frac{\Theta; \vec{s} : \Gamma, t : A \qquad \Theta'; \vec{r} : \Gamma', t' : \sim A}{t \perp t', \Theta, \Theta'; \vec{s} : \Gamma, \vec{r} : \Gamma'}
$$

$$
\frac{\Theta; \vec{s} : \Gamma, t : A \qquad \Theta'; \vec{r} : \Gamma', t' : B}{\Theta, \Theta'; \vec{s} : \Gamma, \vec{r} : \Gamma', t \star t' : A \star B} \qquad \frac{\Theta; \vec{s} : \Gamma, t : A, t' : B}{\Theta; \vec{s} : \Gamma, t + t' : A + B}
$$

Proof expressions may be *reduced* by applying reduction rules:

$$
\begin{aligned}
\Theta, t \perp x, x \perp t' \,; \Delta \quad &\triangleright \quad \Theta, t \perp t' \,; \Delta \; \text{(Communication)} \\
\Theta, t \star s \perp t' + s' \,; \Delta \quad &\triangleright \quad \Theta, t \perp t', s \perp s' \,; \Gamma \; \text{(Contraction)} \\
\Theta, x \perp s; \vec{t} : \Gamma \quad &\triangleright \quad \Theta; \vec{t}[x/s] : \Gamma \; \text{(Cleanup)}
\end{aligned}
$$

Proof expressions are strongly normalisable with unique result. By this device, the symmetry of classical linear logic may be combined with the usual conceptual apparatus for evaluating expressions in term calculi. Proof expressions also offer a neat notation for Girard's "boxes" (Girard 1987).

19 Multiplicative operators

In this chapter we explore the possibility of generalizing the notion of multiplicative operator, or simply "multiplicative" beyond the operators definable from ⊔ and ⋆. Our exposition is based on Danos & Regnier 1989.

19.1. The input-output interpretation of proofnets

The following suggestive informal interpretation may serve as motivation for the search of logical operators permitting cut elimination. The interpretation is neither "logical" nor purely type-theoretical, but is rather like a theory of networks with switches used for the transmission of information.

If we want to transmit information, types (= formulas) serve to guarantee that an output must be of the type requested by the input (Girard uses the terminology of "question" and "answer" in this connection.) In a symmetrical treatment, the roles of input and output may be reversed.

Logical connectives represent ways of "grouping together" information of various types. Thus $A \star B$ represents information obtained by combining information of type A with information of type B.

The simplest correct connection between input and output for primitive types (represented by prime formulas) is given by the axiom ($-$link): *input* of type A matches *output* of type A (symbolized by

$\sim\!A)$ – or input of type $\sim\!A$ matches output of type $\sim\!\sim\!A \equiv A$.

A use of Cut means that two routes of information are being connected: input A flows to output B, which is the same as input $\sim\!B$ and flows to output C, which yields a route from input A to output C, by plugging output B into input $\sim\!B$ (Cut).

An information route should not be short-circuited: output $\sim\!A$ from input A should not be fed back as input, so

$$\frac{A \qquad \sim\!A}{} \quad \text{cut}$$

represents a forbidden plugging.

How do we guarantee that application of a cut to output A and input $\sim\!A$ is a correct way of plugging for compound A (say of the form $A \equiv B\!\star\!C$)? By showing that plugging $B\!\star\!C$ into $\sim\!(B\!\star\!C) \equiv \sim\!B + \sim\!C$ reduces to plugging B into $\sim\!B$, C into $\sim\!C$: cut elimination.

19.2. Multiplicative operators in the sequent calculus

We shall first discuss the notion of a multiplicative operator in the one-sided sequent calculus. We consider here multiplicative operators C specified by a set of introduction rules of the form:

$$\frac{\Gamma_1, A_1, \ldots, A_{i(1)} \quad \Gamma_2, A_{i(1)+1}, \ldots, A_{i(2)} \quad \cdots \quad \Gamma_n, A_{i(n-1)+1}, \ldots, A_{i(n)}}{\Gamma_1, \Gamma_2, \ldots, \Gamma_n, C(A_1, \ldots, A_{i(n)})}$$

Since the $\Gamma_1, \Gamma_2, \ldots, \Gamma_n$ are simply united in the conclusion, the rule is entirely determined by the partition

$$\{1, \ldots, i(1)\}, \ \{i(1)+1, \ldots, i(2)\}, \ldots, \{i(n-1)+1, \ldots, i(n)\}$$

of $\{1, \ldots, i(n)\}$. Thus the introduction rules for an operator of n arguments may be represented by a finite set of partitions of $1, 2, \ldots, n$.

However, this is not sufficient to characterize an operator C: we also need a dual C* with the same number of arguments, permitting us to extend the de Morgan duality by

$$\sim\!C(A_1, \ldots, A_n) := C^*(\sim\!A_1, \ldots, \sim\!A_n).$$

In a two-sided calculus the rules for C* simply correspond to the left rules for C. With respect to the pair C, C* we may ask whether cut elimination remains possible when C, C* are added to the calculus. Crucial for extending cut elimination is the possibility of reducing a cut of the form

$$\frac{\overset{\mathcal{D}}{C(1,2,\ldots,n),\Gamma} \qquad \overset{\mathcal{D}'}{C^*(\sim 1,\sim 2,\ldots,\sim n),\Delta}}{\text{cut}}$$

where the last rule in $\mathcal{D}$, $\mathcal{D}'$ introduced the occurrence of C, C^* shown, to n cuts between $1, \sim 1;\ 2, \sim 2; \ldots n, \sim n$. (Here as in the sequel we often simplify notation by taking simply numbers for the formulas.)

19.3.   DEFINITION. Let p and q be partitions of $\{1,\ldots,n\}$, then $\text{Graph}(p,q)$ is a graph with as nodes the elements of $p \cup q$; for every $k \in p_i \cap q_j$, $p_i \in p$, $q_j \in q$ we add an edge (which might be denoted by k) between p_i and q_j. We say that p is *ortho* to q (notation $p \perp q$) if $\text{Graph}(p,q)$ is acyclic and connected. □

NOTATION. In describing partitions we shall often save on braces writing $n_1 n_2 \ldots n_p$ for $\{n_1, n_2, \ldots, n_p\}$; thus e.g., $\{\{1,2\},\{3,4\}\}$ becomes $\{12, 34\}$. □

As a first approximation to the notion of a sequential multiplicative connective, we take pairs of operators C, C^* with sets of partitions $\text{Part}(C)$, $\text{Part}(C^*)$. We give some examples.

19.4. EXAMPLES

(a) Consider C, C^* with

$$\text{Part}(C) = \{\{13, 24\}, \{12, 34\}\}$$

$$\text{Part}(C^*) = \{\{14, 2, 3\}, \{1, 23, 4\}\}$$

The various graphs $\text{Graph}(p,q)$ for $p \in \text{Part}(C), q \in \text{Part}(C^*)$ take the form

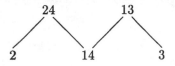 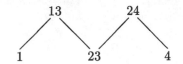

etcetera.

(b) If $\text{Part}(C_1) = \{\{1, 23\}\}$, $\text{Part}(C_1^*) = \{\{123\}\}$ we obtain a graph with a cycle.

(c) With each $\star$, $+$-operator definable by a formula $X(1, 2, \ldots, n)$, where each of $1, 2, \ldots, n$ occurs only once, we can associate a complete set of partitions such that the rule determined by the partition holds for the definable operator, by listing all possible cut free proofs of

$X(1, 2, \ldots, n)$ from sequents containing only the atoms $1, 2, \ldots, n$ as formulas. $(1 \star 2) + (3 \star 4)$ yields the partitions

$$\{\{13, 2, 4\}, \{14, 2, 3\}, \{23, 1, 4\}, \{24, 1, 3\}\}$$

and its dual $(1 + 2) \star (3 + 4)$ has only

$$\{\{12, 34\}\}$$

giving rise to a connective (C_2, C_2^*) with $\mathrm{Part}(C_2) \perp \mathrm{Part}(C_2^*)$.
  (d) $((1 \star 2) + 3) + 4$ yields the partitions

$$\{\{1, 234\}, \{12, 34\}, \{14, 23\}, \{124, 3\}\}$$

and its dual yields
$$\{\{12, 3, 4\}\}$$

giving rise to a connective (C_3, C_3^*) with $\mathrm{Part}(C_3) \perp \mathrm{Part}(C_3^*)$.

19.5. LEMMA. *Cut elimination in a sequent calculus with* (C, C^*) *with rules specified by sets of partitions* $\mathrm{Part}(C)$ *and* $\mathrm{Part}(C^*)$ *extends to cuts involving* C, C^* *with sets of partitions such that* $\mathrm{Part}(C) \perp \mathrm{Part}(C^*)$.

PROOF. $p \perp q$ means that $\mathrm{Graph}(p, q)$ can be successively contracted into a point by taking any edge, identifying its two vertices and removing the edge. Each contraction step, removing an edge m, corresponds to a cut on arguments $m, \sim m$ of $C(1, 2, \ldots, n)$, $C^*(\sim 1, \sim 2, \ldots, \sim n)$. We leave it to the reader to convince himself of this fact in detail. □
  The lemma justifies the following definition.

DEFINITION. An *s-connective (sequential connective)* is a pair of operators C, C^* with sets of partitions $\mathrm{Part}(C)$, $\mathrm{Part}(C^*)$ such that $\mathrm{Part}(C) \perp \mathrm{Part}(C^*)$.
  An s-connective (C, C^*) is *definable* if there is a formula $X(1, 2, \ldots, n)$, in which each argument i occurs only once, defined from $+, \star$ having $\mathrm{Part}(C)$ as the set of partitions induced by possible cut-free derivations of $X(1, 2, \ldots, n)$, while the induced partitions of $\sim X(\sim 1, \sim 2, \ldots, \sim n)$ are precisely $\mathrm{Part}(C^*)$. □
  To reassure ourselves we may prove

LEMMA. *Each* $\star, +$-*formula* $X(1, 2, \ldots, n)$ *in which each argument* i *occurs only once represents an s-connective.*

♠ EXERCISE. Prove the lemma.

19.6. REMARKS

(i) There are undefinable s-connectives; example (a) in 19.4 above is not definable.

(ii) Certain definable connectives with distinct definitions represent the same s-connective, e.g.,

$$\cdot + (\cdot + \cdot) \text{ and } (\cdot + \cdot) + \cdot$$

(iii) Certain sequents provable for definable connectives, with actual use of the definition *defined* connective are not provable from their rules as s-connective.

For example, the sequent $C_2(1, 2, 3, 4), C_3(\sim 1, \sim 3, \sim 2, \sim 4)$, where C_2, C_3 are as in 19.4, is derivable from the definition of C_2, C_3, but not from the rules as given by their partitions. Even non-atomic axioms create a problem: for $C_4(1, 2, 3) := (1 \star 2) + 3$, $C_4^*(1, 2, 3) := (1 + 2) \star 3$ we cannot prove the sequent $C_4(1, 2, 3), C_4^*(\sim 1, \sim 2, \sim 3)$ from the partition rules. (This is called the "packing problem" by Danos & Regnier.)

♠ EXERCISES.

1. Verify the assertions made under (iii) above.
2. Show that (C, C^*) of 19.4 is indeed undefinable.

19.7. Modules

The observation under (iii) above clearly demonstrates the unsatisfactory character of the notion of "s-connective" as a general notion of multiplicative in the sequent calculus. We shall now show that a corresponding generalization of multiplicative operators for proofnets is much more satisfactory. First we consider the notion of a module in the original $\star, +$ fragment.

It is sometimes slightly more convenient to use a dual representation of the graph of a proof structure: each link, and each conclusion is represented by a node, and the formulas are the edges. Thus e.g., the proofnet deriving the axiom $A \star B, \sim A + \sim B$ becomes

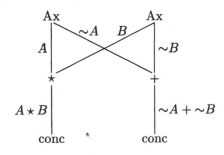

The *switchings* of the preceding chapters correspond here to graphs obtained from a PS-graph by omitting a premise edge for each +-link. Obviously we again have for this dual representation: a proof structure ν is a proofnet iff every switching of ν is acyclic and connected.

19.8. DEFINITION. If X is a subset of the vertices of a PS ν, the *hull* $H_\nu(X)$ of X is obtained as follows: keep all edges connecting elements of X, and for an edge linking $v \in X$ to v' outside X, replace v' by a new node v''; these new nodes are the *border* of the hull.

A *module* in ν is a graph $H_\nu(X)$ with non-empty border generated by a subset of nodes X for which every switching is acyclic. (We do not require a module to be connected.) With every module with border say $\{1, 2, 3, \ldots, n\}$ we may associate the set of partitions of $\{1, 2, \ldots, n\}$ induced by all possible switchings of the module; this is called the *type* of the module. (Each switching determines a partition with as elements the groups of premise border nodes which are connected for the switching.) □

REMARK. In the original definition of a PS as a graph, a module is obtained by taking a set of links, forming the graph with as edges the edges of the links, and as nodes formulas which are common endpoint of two link edges, plus the border nodes. The border nodes correspond to the formulas which are endpoint of an edge and are either terminal or endpoint of another edge not belonging to the module.

EXAMPLE. The proofnet corresponding to the proof of a non-atomic axiom $A, \sim A$ is constructed from two modules, one with a terminal conclusion node below the edge representing A, the other with terminal node with adjacent edge $\sim A$; their premise border nodes correspond to the axioms, and the two modules have the axioms as common border nodes. The module with conclusion A is called the *formula tree* for A; as one easily sees, the formula tree looks in general almost the same as the formula tree for A in the dual representation of proof structures used in the preceding chapters. The picture below gives the example for the case $A \equiv (P \star Q) + R$.

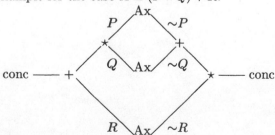

19.9. DEFINITION. Two modules are *connectable* if we can pairwise identify their border nodes, such that the resulting proof structure is a proofnet. □

PROPOSITION. *Two modules are connectable if their respective types are ortho.*

PROOF. Choose an switching of the combined structure; this determines two partitions of the border, say p_1, p_2. Retract in the switching the internal edges of the two modules; we end with Graph(p_1, p_2). □

19.10. Generalized connectives in proof structures

DEFINITION. An n-ary p-*connective* (C,C*) is given by two sets of partitions P,Q over $\{1, 2, \ldots, n\}$ such that $P \perp Q, P^{\perp} \perp Q^{\perp}$.

A PS involving (C,C*) (generalized PS) is constructed as before in Chapter 17, but we have an extra clause: if $\nu, A_1, \ldots, A_n$ is a PS with terminal nodes $A_1, \ldots, A_n$, then ν with edges

$$A_i \relbar\joinrel\relbar C(A_1, \ldots, A_n)$$

added (C-link) is a PS; the new terminal nodes are

$$(\text{TN}(\nu) \cup \{C(A_1, \ldots, A_n)\}) \setminus \{A_1, \ldots, A_n\};$$

similarly for C* (C*-link).

A *switching* of a C-link consists in the choice of a $p \in$ P, $p = \{p_1, \ldots, p_n\}$ say; choose some p_i, choose an edge belonging to p_i, delete the other edges; for all p_j ($j \neq i$) delete all edges belonging to the corresponding border nodes, and connect the border nodes of each p_j with each other without creating a cycle. (N.B. On the dual representation the definition looks virtually the same.)

We now define a proofnet as a generalized PS satisfying the usual condition:

A *proofnet* is a generalized PS such that every switching is acyclic and connected. □

EXAMPLE.

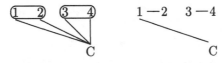

The left side of the picture represents a C-link, with the ovals indicating the classes of a partition; the right side indicates a switching of the link.

19.11. LEMMA. *A proofnet remains a proofnet after performing a cut elimination step.*

PROOF. Take a proofnet with a cut link connecting an occurrence of C and an occurrence of C*; The picture below illustrates (part of) such a proofnet; the ovals indicate classes of partitions, the vertical dots parts of the structure not shown; the dashed rectangle is not part of the structure but plays a role in illustrating the argument.

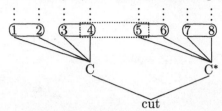

In this proofnet we consider two modules M_1, M_2; M_1 (cf. the picture) consists of the C-link, the C*-link and the cut, and M_2 consists of the rest of the PS. An arbitrary partition r in the type of M_1 is obtained as follows: choose $p \in P$, and $q \in Q$, then

$$r := \{p_i : i \neq i'\} \cup \{q_j : j \neq j'\} \cup \{p_{i'} \cup q_{j'}\}$$

(where $\mathrm{Type}(C) = \{p_1, \ldots, p_n\}$ and $\mathrm{Type}(C^*) = \{q_1, \ldots, q_n\}$) for some choice of i', j'. From this it follows that $s \in \mathrm{Type}(M_2)$ is of the form $s_1 \cup s_2$, where s_1 is a partition over the border corresponding to the C-link, and s_2 a partition over the part of the border corresponding to C*-link. For if not, the original proof structure would not be a proofnet (consider e.g., the dashed rectangle in the picture and assume this to represent an element of the partition of the type of M_2; one sees immediately that then some switching of the proofnet would contain a cycle). Also $s_1 \perp p$ and $s_2 \perp q$, hence by the definition of a p-connective $s_1 \perp s_2$, and so the resulting proof structure after one reduction step is again a proofnet by the same argument used above for the lemma on cut elimination for s-connectives. □

19.12. Definable connectives in proofnets

Finally we want to show that the definable connectives may be regarded as special instances of p-connectives. With an occurrence $X(A_1, \ldots, A_n)$ in a PS of a definable connective represented by the formula $X(1, 2, \ldots, n)$ with 1, 2, ..., n atomic, we can associate the module generated by the nodes representing the connective; there are border nodes corresponding to the arguments ("premises") A_1, ..., A_n and a single border node corresponding to $X(A_1, \ldots, A_n)$ itself.

This module is nothing but the formula tree of $X(1, 2, \ldots, n)$ with $A_1, \ldots, A_n$ substituted for $1, \ldots, n$. Such a module is clearly connected and acyclic. The switchings of this module determine on the premise border nodes a set of partitions P.

Contracting all interior edges not having a border node or terminal node as a vertex transforms the module into a C_X-link, where (C_X, C_X^*) is to become the p-connective corresponding to X. In the sequel we shall not always bother to distinguish notationally between X and its dual on the one hand, and (C_X, C_X^*) on the other hand. Similarly for $\sim X(1, \ldots, n)$ yielding a set of partitions Q; (P,Q) is the set of partitions describing X as a p-connective.

PROPOSITION. *A definable connective is a p-connective.*

PROOF. Let the connective C_X, C_X^* represented by formula X have a pair (P,Q) as characteristic partitions, and consider a proof of $X + \sim X$. Take the formula tree of X and the formula tree of $\sim X$; these represent two modules, with as border a set of nodes corresponding to the axioms, one node for each axiom. Connecting these modules yields the proofnet for the axiom $X + \sim X$. If we choose any switching of this PS, this amounts to choosing an switching in the module of the formula tree for X and choosing a switching in the module of the formula tree of $\sim X$. Contracting the inner edges of the two switchings leaves us obviously with some $\text{Graph}(p, q)$ for a $p \in P$, $q \in Q$. Secondly we have to check that $P^\perp \perp Q^\perp$.

For any partition p, there is a definable X_p with first set of partitions $\{p\}$: if the partition is $\{\{i_{1,1}, \ldots, i_{1,n_1}\}, \{i_{2,1}, \ldots, i_{2,n_2}\}, \ldots, \{i_{p,1}, \ldots, i_{p,n_p}\}\}$ then we take

$$(i_{1,1} \star \cdots \star i_{1,n_1}) + \cdots + (i_{p,1} \star \cdots \star i_{p,n_p})$$

Now choose s_1 in $P^\perp$ and link the formula tree of X and the formula tree of X_{s_1} by identifying the nodes corresponding to axioms.

Do the same for s_2 in $Q^\perp$ and link the two proofnets with a cut applied to X and $\sim X$. Normalization produces a proofnet, and so $s_1 \perp s_2$. The situation is illustrated in the picture below.

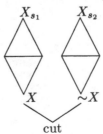

cut

In this picture the horizontal lines represent the identified border nodes of the formula tree modules corresponding to axiom links. □

19.13. Further results

(i) As to the relation between $(\mathrm{Part}(C_X), \mathrm{Part}(C_X^*))$ (the definable X as an s-connective) and (P_X, Q_X) (the definable X as a p-connective), Danos and Regnier prove

$$\mathrm{Part}(C_X^*) \supset P_X, \quad \mathrm{Part}(C_X) = P_X^\perp.$$

(ii) With a permutation σ of $\{1, 2, \ldots, n\}$ we can associate $\mathrm{Part}(\sigma)$, the partition with as its elements the sets of elements of the cycles of σ.

For permutations σ, τ such that $\sigma \perp \tau$ we define the *communication sequence* as

$$1, \sigma(1), \tau\sigma(1), \sigma\tau\sigma(1) \ldots, \sigma(\tau\sigma)^{n-1}(1).$$

This sequence is non-interlacing if the configuration $\ldots i \ldots j \ldots i \ldots j \ldots$ does not occur. Danos & Regnier show:

$\sigma \perp \tau$ and the communication sequence is non-interlacing
$\Leftrightarrow \mathrm{Part}(\sigma) \perp \mathrm{Part}(\tau)$.

♠ EXERCISE. Prove (ii).

20   The undecidability of linear logic

20.1.   In this chapter we show the undecidability of linear logic,
following Lincoln et al. 1990a. The idea is to simulate the action
of (non-deterministic) two-countermachines (two-register machines)
in linear logic; the acceptance problem for two-counter machines is
known to be undecidable (see e.g., Minsky 1961). Since a simula-
tion of zero-test instructions of two-counter machines is inconvenient,
we replace the two-counter machines by a variant, the *and-branching*
two-counter machines, or ACM's. The proof is easily adapted to
n-counter machines and shows in fact much more than just undecidability: machine computations can be faithfully represented in linear
logic.

20.2. DEFINITION.   A *two-counter machine* M has two counters a, b,
a finite set of states $q_1, \ldots, q_n$, among which an *initial state* q_I, and a
subset Fin of *final* states.

An *instantaneous description* (ID) of the machine is a triple $\langle q_i, x,$
$y \rangle$, where x, y are the values of a and b respectively, and q_i a state of
the machine. The instructions of M are of the following six types:

instruction	transition of ID's
$q_i + aq_j$	$\langle q_i, x, y \rangle \mapsto \langle q_j, x+1, y \rangle$
$q_i + bq_j$	$\langle q_i, x, y \rangle \mapsto \langle q_j, x, y+1 \rangle$
$q_i - aq_j$	$\langle q_i, x+1, y \rangle \mapsto \langle q_j, x, y \rangle$
$q_i - bq_j$	$\langle q_i, x, y+1 \rangle \mapsto \langle q_j, x, y \rangle$
$q_i 0 aq_j$	$\langle q_i, 0, y \rangle \mapsto \langle q_j, 0, y \rangle$
$q_i 0 bq_j$	$\langle q_i, x, 0 \rangle \mapsto \langle q_j, x, 0 \rangle$

The first two instructions are the $+$-*instructions*, the next two are the $-$-*instructions*, and the last two are the *zero-tests*. M *accepts* $\langle q_i, x, y \rangle$ if there is a sequence of transitions ending in $\langle q_i, 0, 0 \rangle$ with $q_i \in$ Fin.
$\square$

REMARK. Without loss of generality we may assume that M has a single terminal state q_F for which no outgoing transition exists (no instruction begins with q_F).

To see this, we transform a given machine M into an M$'$ as follows: add two new states q_D, q_F, and add instructions for each $q_j \in$ Fin:

$$q_j + a q_D \qquad q_D - a q_F$$

It is easy to see that M$'$ accepts the same ID's as M does. Henceforth we shall assume our two-countermachines to have such a unique q_F without outgoing transitions.

20.3. DEFINITION. An *ACM* (*and-branching two-counter machine*) M has states $q_1, \ldots, q_n$ among which a unique initial state q_I and a unique final state q_F without outgoing transitions, and two counters a, b. An *instantaneous description* (*ID*) is a finite list of triples $\langle q_i, x, y \rangle$. The instructions are $+$-instructions, $-$-instructions (as for two-counter machines), and *forking instructions*:

instruction transition
$q_i f q_j q_k \qquad \ldots \langle q_i, x, y \rangle \ldots \mapsto \ldots \langle q_j, x, y \rangle, \langle q_k, x, y \rangle \ldots$

An ID is accepted by M if there is a finite sequence of transitions ending in a sequence consisting of triples $\langle q_F, 0, 0 \rangle$ only. $\square$

20.4. LEMMA. *To each two-counter machine* M *we can find an ACM* M$'$ *such that* M$'$ *has all the states of* M, *and the same initial and final state as* M, *and accepts the sequence (with single element)* $\langle q_i, x, y \rangle$ *iff* M *accepts* $\langle q_i, x, y \rangle$.

PROOF. Given M, we construct M$'$ as follows. M$'$ has the same $+$- and $-$-instructions as M. M$'$ has two additional states z_a, z_b (zero test for a, zero test for b respectively) with instructions

$$z_a - b z_a \qquad z_b - a z_b \qquad z_a f q_F q_F \qquad z_b f q_F q_F;$$

instructions $q_i 0 a q_j, q_i 0 b q_j$ of M are replaced by

$$q_i f q_j z_a, \quad q_i f q_j z_b$$

respectively. The effect of this is that when M encounters a zero test $q_i 0 a q_j$ say, which applied to $\langle q_i, x, y \rangle$ yields $\langle q_j, x, y \rangle$ if $a = 0$, then M' at the corresponding step of its computation starts a parallel computation; with $\langle q_j, x, y \rangle$ M' continues as M, but the new triple in the ID $\langle z_a, x, y \rangle$ produced by the instruction $q_i f q_j z_a$ serves to perform a test whether $a = 0$. For we can only accept the triple $\langle z_a, x, y \rangle$: if $a = 0$, $z_a - b z_a$ decreases y until we arrive at $\langle z_a, 0, 0 \rangle$, and then forks into $\langle q_F, 0, 0 \rangle, \langle q_F, 0, 0 \rangle$.

20.5. EXAMPLE. Let M be a two-counter machine with instructions

$$\delta_1 := q_I + a q_2, \quad \delta_2 := q_3 - a q_F, \quad \delta_3 := q_2 0 b q_3$$

A computation of this machine is:

$$\langle q_I, 0, 0 \rangle \overset{\delta_1}{\mapsto} \langle q_2, 1, 0 \rangle \overset{\delta_2}{\mapsto} \langle q_3, 1, 0 \rangle \overset{\delta_3}{\mapsto} \langle q_F, 0, 0 \rangle$$

The corresponding ACM has

$$\delta_1' := \delta_1, \quad \delta_2' := \delta_2, \quad \delta_3' := q_2 f z_b q_3, \quad \delta_4' := z_b - a z_b, \quad \delta_5' := z_b f q_F q_F$$

The computation now becomes

$$\langle q_I, 0, 0 \rangle \overset{\delta_1'}{\mapsto} \langle q_2, 1, 0 \rangle \overset{\delta_3'}{\mapsto} \langle z_b, 1, 0 \rangle \langle q_3, 1, 0 \rangle \overset{\delta_4'}{\mapsto} \langle z_b, 0, 0 \rangle \langle q_3, 1, 0 \rangle \overset{\delta_5'}{\mapsto}$$

$$\langle q_F, 0, 0 \rangle \langle q_F, 0, 0 \rangle \langle q_3, 1, 0 \rangle \overset{\delta_3'}{\mapsto} \langle q_F, 0, 0 \rangle \langle q_F, 0, 0 \rangle \langle q_F, 0, 0 \rangle$$

20.6. Refinement of cut elimination

DEFINITION. We consider theories in $\mathbf{CLL}_0$ without $\sim, \mathbf{0}$, given by finitely many axioms of the form

$$(*) \qquad\qquad P_{i_1}, \ldots, P_{i_k} \Rightarrow C$$

($P_0, P_1, \ldots$ is our supply of propositional variables). Axioms may be used arbitrarily often in deductions. We call a cut involving an axiom $(*)$ with C as cut formula a *principal cut*. The *cut degree* of a derivation is defined as before, except that a principal cut is reckoned to have degree 0. A proof containing only principal cuts is said to be *standard*. $\square$

20.7. LEMMA. *Each deduction in the two-sided calculus for* $\mathbf{CLL}_0$ *without* $\sim, \mathbf{0}$ *can be transformed into a standard deduction by the standard cut elimination procedure.*

♠ EXERCISE. Check the necessary cases. Can you extend the proof to cover ! and ? as well?

20.8. Encoding ACM's into linear logic

Choose for each q_i of the ACM M a distinct propositional variable Q_i. The instructions of M are now translated into axioms of linear logic according to the following table:

instruction	axiom
$q_i + aq_j$	$Q_i \Rightarrow Q_j \star A$
$q_i + bq_j$	$Q_i \Rightarrow Q_j \star B$
$q_i - aq_j$	$Q_i, A \Rightarrow Q_j$
$q_i - bq_j$	$Q_i, B \Rightarrow Q_j$
$q_i f q_j q_k$	$Q_i \Rightarrow Q_j \sqcup Q_k$

The resulting set of axioms is called Theory(M). A triple $\langle q_i, x, y \rangle$ is translated as

$$\theta(\langle q_i, x, y \rangle) := Q_i, A^x, B^y \Rightarrow Q_F$$

where A^x is a multiset of x copies of A etc. Id's are translated as the sequence of translated triples.

Our aim is now to prove the following

THEOREM. *An ACM M accepts ID S iff every sequent in $\theta(S)$ is deducible in* Theory(M).

The proof of this theorem is split into two lemmas, one for each direction of the theorem.

20.9. LEMMA. *If an ACM M accepts ID S, then every sequent in $\theta(S)$ is deducible.*

PROOF. We have to show that acceptance means that for each element of $\theta(S)$ we can give a deduction. The proof is by induction on the length of the sequence of transitions leading to a final ID $\langle q_F, 0, 0 \rangle$.

If there are 0 transitions, each element of S is of the form $\langle q_F, 0, 0 \rangle$; $\theta(\langle q_F, 0, 0 \rangle) = Q_F \Rightarrow Q_F$, which is an axiom. The induction step is by cases on the type of the first transition.

Case 1. The first instruction applied is $q_i + aq_j$, leading from $\langle q_i, x, y \rangle$ to $\langle q_j, x+1, y \rangle$. By IH (induction hypothesis) we have a deduction $\mathcal{D}$ of $Q_j, A^{x+1}, B^y \Rightarrow Q_F$. Then

$$\cfrac{Q_i \Rightarrow Q_j \star A \qquad \cfrac{\cfrac{\mathcal{D}}{Q_j, A, A^x, B^y \Rightarrow Q_F}}{Q_j \star A, A^x, B^y \Rightarrow Q_F}}{Q_i, A^x, B^y \Rightarrow Q_F}$$

is a correct deduction, since the left premise is an axiom.

Case 2. Instruction $q_i + bq_j$ is treated similarly.

Case 3. The instruction $q_i - aq_j$ at the first transition transforms $\langle q_i, x+1, y \rangle$ into $\langle q_j, x, y \rangle$. By IH there is a deduction $\mathcal{D}$ of $Q_j, A^x, B^y \Rightarrow Q_F$, hence the following is correct deduction:

$$
\cfrac{Q_i, A \Rightarrow Q_j \qquad \cfrac{\mathcal{D}}{Q_j, A^x, B^y \Rightarrow Q_F}}{Q_i, A^{x+1}, B^y \Rightarrow Q_F}
$$

Case 4. Instruction $q_i - bq_j$ is treated similarly.

Case 5 The instruction $q_i f q_j q_k$ transforms at the first step $\langle q_i, x, y \rangle$ in the ID into the pair $\langle q_j, x, y \rangle, \langle q_k, x, y \rangle$. Then the following is a correct deduction

$$
\cfrac{Q_i \Rightarrow Q_j \sqcup Q_k \qquad \cfrac{\cfrac{\mathcal{D}}{Q_j, A^x, B^y \Rightarrow Q_F} \quad \cfrac{\mathcal{D}'}{Q_k, A^x, B^y \Rightarrow Q_F}}{Q_j \sqcup Q_k, A^x, B^y \Rightarrow Q_F}}{Q_i, A^x, B^y \Rightarrow Q_F}
$$

where $\mathcal{D}, \mathcal{D}'$ exist by IH. $\square$

20.10. LEMMA. *If every sequent in $\theta(S)$ is derivable in* Theory(M), *then M accepts S.*

PROOF. We assume the set $\theta(S)$ to have a set of standard proofs, and we apply induction on the sum of the lengths (number of applications of rules and axioms) of the standard proofs.

Let $\langle q_i, x, y \rangle \in S$, then the standard proof of $\theta(\langle q_i, x, y \rangle)$ must end with $Q_i, A^x, B^y \Rightarrow Q_F$. The only rules or axioms which are applicable are Ax, Cut, or an axiom of Theory(M).

Case 1. Ax: $x = y = 0$, $Q_i \equiv Q_F$, which encodes the accepting triple $\langle q_F, 0, 0 \rangle$.

Case 2. An axiom of Theory(M) is applied. The only possibilities are axioms $Q_i, A \Rightarrow Q_F$ or $Q_i, B \Rightarrow Q_F$, corresponding to triples $\langle q_i, 1, 0 \rangle$, $\langle q_i, 0, 1 \rangle$ respectively. One application of $q_i - aq_F$ or $q_i - bq_F$ leads to the accepted $\langle q_F, 0, 0 \rangle$.

Case 3. The last step in the proof is Cut. By our cut elimination theorem, the proof must therefore have the form

$$
\cfrac{Q_i, \Gamma' \Rightarrow C \qquad \cfrac{\mathcal{D}}{\Gamma, C \Rightarrow Q_F}}{Q_i, \Gamma', \Gamma \Rightarrow Q_F}
$$

with $x + y \leq 1$. We can distinguish five subcases, according to the axiom involved in the cut.

Case 3a. The axiom is $Q_i \Rightarrow Q_j \star A$ corresponding to an instruction $q_i + aq_j$. The proof takes the form

$$\frac{Q_i \Rightarrow Q_j \star A \qquad \overset{\mathcal{D}}{\Gamma, Q_j \star A \Rightarrow Q_F}}{Q_i, \Gamma \Rightarrow Q_F}$$

The only rule which can be applied to obtain the right premise are L$\star$, Ax, or Cut. But the sequent does not have the form of an axiom of Theory(M), and Ax is also excluded. So Cut remains, but Cut in a standard proof involves an axiom of the theory, with on the left side a Q_k which cannot be a cut formula and hence should appear in the antecedent of the conclusion of the Cut. This is not the case, so the rule applied to get the second premise is an application of L$\star$, so the deduction ended with

$$\frac{Q_i \Rightarrow Q_j \star A \qquad \dfrac{Q_j, A^{x+1}, B^y \Rightarrow Q_F}{Q_j \star A, A^x, B^y \Rightarrow Q_F}}{Q_i, A^x, B^y \Rightarrow Q_F}$$

and since $Q_i, A^{x+1}, B^y \Rightarrow Q_F \equiv \theta(\langle q_i, x+1, y \rangle)$ with a smaller proof, $\langle q_i, x+1, y \rangle$ is accepted, and hence also $\langle q_i, x, y \rangle$ is accepted.

Case 3b. The axiom involved in the cut is $Q_i \Rightarrow Q_j \sqcup Q_k$. Hence the proof must end

$$\frac{Q_i \Rightarrow Q_j \sqcup Q_k \qquad \overset{\mathcal{D}}{Q_j \sqcup Q_k, A^x, B^y \Rightarrow Q_F}}{Q_i, A^x, B^y \Rightarrow Q_F}$$

The last rule in $\mathcal{D}$ must be either L$\sqcup$, or Cut, or Ax, or an element of Theory(M). Ax and elements of Theory(M) are excluded as before; if the last rule in $\mathcal{D}$ had been Cut, the right hand premise should contain some Q_l in the antecedent which is not the case. Hence $\mathcal{D}$ ends with

$$\frac{Q_j, A^x, B^y \Rightarrow Q_F \qquad Q_k, A^x, B^y \Rightarrow Q_F}{Q_j \sqcup Q_k, A^x, B^y \Rightarrow Q_F}$$

and we can again apply the induction hypothesis. The other subcases are left to the reader. $\square$ We have now shown that

20.11. PROPOSITION. *Derivability of sequents in* $\mathbf{T} + \mathbf{ILL}_0$ *or* $\mathbf{T} + \mathbf{CLL}_0$ *is for arbitrary finite theories* $\mathbf{T}$ *undecidable.*

N.B. The preceding argument works for both cases since we need to consider only intuitionistic proofs, as we see by inspection.

LEMMA. *Let* $\mathbf{T} \equiv \{T_1, \ldots, T_n\}$,

$$T_i \equiv P_{i,1}, \ldots, P_{i,n(i)} \Rightarrow C_i$$

and let

$$[T_i] := !(P_{i,1} \star \cdots \star P_{i,n(i)} \multimap C_i)$$

then $\Gamma \Rightarrow \Delta$ *is derivable in* $\mathbf{T}+\mathbf{CLL}$ *($\mathbf{T}+\mathbf{ILL}$) iff* $\Gamma, [T_1], \ldots [T_n] \Rightarrow \Delta$ *is derivable in* $\mathbf{CLL}$ *($\mathbf{ILL}$).*

EXERCISE. Prove the lemma.

20.12. THEOREM. $\mathbf{CLL}_e$ *and* $\mathbf{ILL}_e$ *are undecidable.*

PROOF. Immediate by the preceding lemma and proposition.

REMARK. The whole proof is easily adapted to the use of machines with an arbitrary finite number of counters. In this case we can rely on the unsolvability of the acceptance problem for such machines, which is easily proved; see e.g., Lambek 1961.

21 Cut elimination and strong normalization

The following chapter has been contributed by D. Roorda and constitutes part of Roorda 1991. It contains a proof of "strong normalization" for cut elimination; the method is taken from Dragalin 1988, appendix B, where it is applied to Gentzen's systems.

21.1. Preliminaries

In an application of a rule, the formulas that match the Γ, Γ_1, Γ_2, Δ, Δ_1, Δ_2 are called *side formulas*, the others that occur in the conclusion *major formulas*, and the others that occur in the premises *minor formulas*. In this chapter we drop commas between parts of a multiset, i.e., we write $\Gamma\Delta$ for Γ, Δ.

Cuts on the multiplicative constants (i.e., **1** and **0**) have a rather trivial nature. They are easy to remove, and we do not want to bother about them after the following lemma.

Suppose a constant **1** or **0** is introduced in a derivation, and it is not used as component for bigger formulas. Then we can draw a line through its occurrences, such that occurrences in premises are connected to occurrences in conclusions. Sometimes lines meet (in case of additive, binary rules) sometimes there is a choice (when there are several constants in the conclusion). Anyway, to each occurrence of such a constant in the conclusion of a derivation, we can associate a tree of such connecting lines. An easy inspection shows that such trees can be removed from a derivation, without destroying validity of the applications of the inference rules: those constants must have been introduced by **L1** or **R0**, so let us remove those introductions. The rest of the occurrences are merely side formulas, so it is harmless to remove them.

21.2. LEMMA. *Every proof of a sequent* $\Gamma 1 \vdash \Delta$ *(resp.* $\Gamma \vdash 0\Delta$*) can be transformed into a proof of* $\Gamma \vdash \Delta$ *with exactly the same structure, but with the difference that all occurrences of* **1** *(resp.* **0***) that are connected to the occurrence in the conclusion, are removed.*

21.3. Case analysis of cut applications

Consider the two premises

$$(l) \quad \Gamma_1 \vdash A^n \Delta_1 \qquad (r) \quad \Gamma_2 A^m \vdash \Delta_2$$

of a cut application. (Of course, the n and m are not completely general. Both are non-zero, at most one of them is > 1, and $n > 1$ implies $A \equiv ?B$ and $m > 1$ implies $A \equiv !B$.) The process of eliminating such an application of cut has two key steps: if the cut formula has just been introduced by a logical rule *on both sides* then we can break up the formula, and replace the cut by two cuts on the immediate subformulas of the original one. If this principal case ((5) in the list below) does not apply, then it appears to be possible to permute the cut with previous rules, which is a positive step towards a future principal case. We distinguish between *logical* rules, and rules introducing the modalities ! and ?. These formulas create spots where the structural rules, among which contraction, are permissible. Extra care is therefore needed, which shows in a considerable proliferation of cases. The following list of cases and subcases is complete; Cases (1), (2), and (5) are perfectly standard, but Cases (3) and (4) exhibit the peculiarities of linear logic.

Case (1) (l) or (r) is an axiom of the form $A \vdash A$; so $n = m = 1$

Case (2) (l) is $\vdash \mathbf{1}$ or (r) is $\mathbf{0} \vdash$; so $n = m = 1$

Case (3)   In at least one of (l) and (r) all occurrences of A involved in the cut are not major formulas; the last rule applied there is:

   a. $\perp$ or $\top$;

   b. a logical rule with one premise (this excludes R! and L?);

   c. a parallel logical rule with two premises: R$\star$, L+, L$\multimap$;

   d. Cut;

   e. a sequential logical rule with two premises: R$\sqcap$, L$\sqcup$;

   f. R! or L?

Case (4)   In both (l) and (r) A is major formula; in at least one of (l) and (r) the last rule applied is

   a. W? or W!

   b. R? or L!

   c. C? or C!

Case (5)   In both (l) and (r) A is major formula; and in both A is introduced according its principal connective $\star$, +, $\sqcap$, $\sqcup$, $\multimap$, or $\sim$. So $n = m = 1$.

21.4. Primitive reductions

According the distinctions above we shall give reductions of proofs that end with a cut application. If there are symmetrical cases, we treat only one representative.

(1)

$$\frac{A \vdash A \quad \Gamma_2 A \vdash \Delta_2}{\Gamma_2 A \vdash \Delta_2} \text{ cut} \quad \rightsquigarrow \quad \Gamma_2 A \vdash \Delta_2$$

(2)

$$\frac{\vdash 1 \quad \Gamma_2 1 \vdash \Delta_2}{\Gamma \vdash \Delta} \text{ cut} \quad \rightsquigarrow \quad \Gamma \vdash \Delta \qquad \text{by Lemma 21.2}$$

(3a)

$$\frac{\Gamma_1 \perp \vdash A^n \Delta_1 \quad \Gamma_2 A^m \vdash \Delta_2}{\Gamma_1 \Gamma_2 \perp \vdash \Delta_1 \Delta_2} \text{ cut} \quad \rightsquigarrow \quad \Gamma_1 \Gamma_2 \perp \vdash \Delta_1 \Delta_2$$

(3b)

$$\frac{\dfrac{\vdots}{\Gamma \vdash A^n \Delta} \text{ rule}}{\dfrac{\Gamma_1 \vdash A^n \Delta_1 \qquad \dfrac{\vdots}{\Gamma_2 A^m \vdash \Delta_2}}{\Gamma_1 \Gamma_2 \vdash \Delta_1 \Delta_2}} \text{ cut} \qquad \rightsquigarrow \qquad \frac{\dfrac{\vdots}{\Gamma \vdash A^n \Delta} \quad \dfrac{\vdots}{\Gamma_2 A^m \vdash \Delta_2}}{\dfrac{\Gamma \Gamma_2 \vdash \Delta \Delta_2}{\Gamma_1 \Gamma_2 \vdash \Delta_1 \Delta_2}} \text{ rule}$$

(3c)   There are several subcases; the simplest occurs when all occurrences of A that are involved in the cut come from only one premise of the multiplicative logical rule:

$$\frac{\dfrac{\dfrac{\vdots}{\Gamma_1^1 \vdash C A^n \Delta_1^1} \quad \dfrac{\vdots}{\Gamma_1^2 \vdash D \Delta_1^2}}{\Gamma_1^1 \Gamma_1^2 \vdash C \circ D \, A^n \Delta_1^1 \Delta_1^2} \text{ rule} \quad \dfrac{\vdots}{\Gamma_2 A^m \vdash \Delta_2}}{\Gamma_1^1 \Gamma_1^2 \Gamma_2 \vdash C \circ D \, \Delta_1^1 \Delta_1^2 \Delta_2} \text{ cut} \qquad \rightsquigarrow$$

$$\frac{\dfrac{\dfrac{\vdots}{\Gamma_1^1 \vdash C A^n \Delta_1^1} \quad \dfrac{\vdots}{\Gamma_2 A^m \vdash \Delta_2}}{\Gamma_1^1 \Gamma_2 \vdash C \Delta_1^1 \Delta_2} \text{ cut} \quad \dfrac{\vdots}{\Gamma_1^2 \vdash D \Delta_1^2}}{\Gamma_1^1 \Gamma_1^2 \Gamma_2 \vdash C \circ D \, \Delta_1^1 \Delta_1^2 \Delta_2} \text{ rule}$$

For this reduction it is immaterial whether the C, D, and $C \circ D$ occur left or right. The other case is typically like this:

$$\frac{\dfrac{\dfrac{\vdots}{\Gamma_1^1 \vdash C(?A)^k \Delta_1^1} \quad \dfrac{\vdots}{\Gamma_1^2 \vdash D(?A)^l \Delta_1^2}}{\Gamma_1^1 \Gamma_1^2 \vdash C \circ D \, (?A)^{k+l} \Delta_1^1 \Delta_1^2} \text{ rule} \quad \dfrac{\vdots}{\Gamma_2 ?A \vdash \Delta_2}}{\Gamma_1^1 \Gamma_1^2 \Gamma_2 \vdash C \circ D \, \Delta_1^1 \Delta_1^2 \Delta_2} \text{ cut} \qquad \rightsquigarrow \cdots$$

Now we are forced to permute on the other premise. But that could be a problem in three cases:

(i)   The same phenomenon occurs in the other premise. Impossible, for only one occurrence of $?A$ at this side can be involved in the cut.

(ii)   The second premise ends with a cut. Then we do not provide any reduction for the cut of the conclusion, but there is at least one other cut to apply a reduction to.

(iii)   In the second premise $?A$ was just introduced. But then that premise is of the form $!\Gamma_2\,?A \vdash ?\Delta_2$ so that we can proceed in the following way: cut the right premise with both premises of the logical rule, apply that rule on the results, and finish with a sequence of $?$- and $!$- contractions.

(3d)   In this case we do not provide a reduction, but there is another cut to apply a reduction to.

(3e)

$$\frac{\dfrac{\vdots \qquad \vdots}{\dfrac{\Gamma_1 \vdash CA^n\Delta_1 \quad \Gamma_1 \vdash DA^n\Delta_1}{\Gamma_1 \vdash C \circ D\ A^n\Delta_1}\ \text{rule} \qquad \dfrac{\vdots}{\Gamma_2 A^m \vdash \Delta_2}}}{\Gamma_1\Gamma_2 \vdash C \circ D\ \Delta_1\Delta_2}\ \text{cut} \qquad \rightsquigarrow$$

$$\frac{\dfrac{\dfrac{\vdots \qquad \vdots}{\Gamma_1 \vdash CA^n\Delta_1 \quad \Gamma_2 A^m \vdash \Delta_2}}{\Gamma_1\Gamma_2 \vdash C\Delta_1\Delta_2}\ \text{cut} \quad \dfrac{\dfrac{\vdots \qquad \vdots}{\Gamma_1 \vdash DA^n\Delta_1 \quad \Gamma_2 A^m \vdash \Delta_2}}{\Gamma_1\Gamma_2 \vdash D\Delta_1\Delta_2}\ \text{cut}}{\Gamma_1\Gamma_2 \vdash C \circ D\ \Delta_1\Delta_2}\ \text{rule}$$

For this reduction it is immaterial whether the C, D, and $C \circ D$ occur left or right. $\circ$ stands for $\sqcap$ or $\sqcup$.

(3f)   We consider only one typical case.   Consider $\Gamma_1 \vdash (A)^n \Delta_1$. Suppose that all the indicated occurrences of A are side formulas. If R! is applied we have the situation

$$\frac{\dfrac{\vdots}{!\Gamma_1 \vdash (?A)^n?\Delta_1 C}}{\dfrac{!\Gamma_1 \vdash (?A)^n?\Delta_1 !C \quad \Gamma_2?A \vdash \Delta_2}{!\Gamma_1\Gamma_2 \vdash ?\Delta_1\Delta_2 !C}}\ \text{cut}$$

and if L? is applied, we have the similar situation

$$\frac{\dfrac{\vdots}{C!\Gamma_1 \vdash (?A)^n?\Delta_1}}{\dfrac{?C!\Gamma_1 \vdash (?A)^n?\Delta_1 \quad \Gamma_2?A \vdash \Delta_2}{?C!\Gamma_1\Gamma_2 \vdash ?\Delta_1\Delta_2}}\ \text{cut}$$

If we try to permute the cut with the L? then we encounter the problem that after the cut we may not have a good premise for L?. So, in this case we are forced to permute on the other premise. But that could be a problem in three cases:

(i)   The situation in the second premise is the mirror image of the first premise. But then $?A$ should begin with a !, which is not so.

(ii)   The second premise ends with a cut. Then we do not provide any reduction for this cut, but there is at least one other cut to apply a reduction to.

(iii) In the second premise $?A$ was just introduced. But then we have a situation in which it is possible to permute on the left premise! (we show only the last situation):

$$\frac{\dfrac{\vdots}{C!\Gamma_1 \vdash (?A)^n ?\Delta_1} \quad \dfrac{\vdots}{!\Gamma_2 A \vdash ?\Delta_2}}{\dfrac{?C!\Gamma_1 \vdash (?A)^n ?\Delta_1 \quad !\Gamma_2 ?A \vdash ?\Delta_2}{?C!\Gamma_1 !\Gamma_2 \vdash ?\Delta_1 ?\Delta_2} \text{ cut}} \quad \leadsto$$

$$\frac{\dfrac{\dfrac{\vdots}{C!\Gamma_1 \vdash (?A)^n ?\Delta_1} \quad \dfrac{\vdots}{!\Gamma_2 ?A \vdash ?\Delta_2}}{C!\Gamma_1 !\Gamma_2 \vdash ?\Delta_1 ?\Delta_2} \text{ cut}}{?C!\Gamma_1 !\Gamma_2 \vdash ?\Delta_1 ?\Delta_2}$$

(4a)   We have W? in (l) or W! in (r). Let us treat the first possibility. Note, that in (r) $?A$ was just introduced.

Situation (1) $?A$ occurs in Δ_1 and such an occurrence is involved in the cut:

$$\frac{\dfrac{\dfrac{\vdots}{\Gamma_1 \vdash (?A)^{n+1}\Delta_1}}{\Gamma_1 \vdash ?A(?A)^{n+1}\Delta_1} \quad \dfrac{\vdots}{!\Gamma_2 ?A \vdash ?\Delta_2}}{\Gamma_1 !\Gamma_2 \vdash \Delta_1 ?\Delta_2} \text{ cut} \quad \leadsto$$

$$\frac{\dfrac{\vdots}{\Gamma_1 \vdash (?A)^{n+1}\Delta_1} \quad \dfrac{\vdots}{!\Gamma_2 ?A \vdash ?\Delta_2}}{\Gamma_1 !\Gamma_2 \vdash \Delta_1 ?\Delta_2} \text{ cut}$$

Situation (2) Otherwise we have the following situation:

$$\frac{\dfrac{\dfrac{\vdots}{\Gamma_1 \vdash \Delta_1}}{\Gamma_1 \vdash ?A\Delta_1} \quad \dfrac{\vdots}{!\Gamma_2 ?A \vdash ?\Delta_2}}{\Gamma_1 !\Gamma_2 \vdash \Delta_1 ?\Delta_2} \text{ cut} \quad \leadsto \quad \frac{\dfrac{\vdots}{\Gamma_1 \vdash \Delta_1}}{\vdots \; W! \; \vdots \; W? \; \vdots} \atop {\Gamma_1 !\Gamma_2 \vdash \Delta_1 ?\Delta_2}$$

(4b)   We have R? in (l) or L! in (r). Note, that in (r) $?A$ was just introduced.

Situation (1) $?A$ occurs in Δ_1 and such an occurrence is involved in the cut:

$$\frac{\dfrac{\dfrac{\vdots}{\Gamma_1 \vdash A(?A)^{n+1}\Delta_1}}{\Gamma_1 \vdash ?A(?A)^{n+1}\Delta_1} \quad \dfrac{\vdots}{!\Gamma_2 A \vdash ?\Delta_2}}{\Gamma_1 !\Gamma_2 \vdash \Delta_1 ?\Delta_2} \text{ cut} \quad \leadsto$$

$$\dfrac{\dfrac{\vdots\qquad\qquad\vdots}{\Gamma_1 \vdash A(?A)^{n+1}\Delta_1 \quad !\Gamma_2?A \vdash ?\Delta_2}\;\text{cut}\qquad \vdots}{\dfrac{\Gamma_1!\Gamma_2 \vdash A\Delta_1?\Delta_2 \qquad\qquad !\Gamma_2 A \vdash ?\Delta_2}{\Gamma_1!\Gamma_2!\Gamma_2 \vdash \Delta_1?\Delta_2?\Delta_2}\;\text{cut}}$$

$$\dfrac{\vdots\; C!\;\vdots\; C?\;\vdots}{\Gamma_1!\Gamma_2 \vdash \Delta_1?\Delta_2}$$

Situation (2) Otherwise we have the following situation:

$$\dfrac{\dfrac{\vdots}{\Gamma_1 \vdash A\Delta_1}\quad \dfrac{\vdots}{!\Gamma_2 A \vdash ?\Delta_2}}{\dfrac{\Gamma_1 \vdash ?A\Delta_1 \quad !\Gamma_2?A \vdash ?\Delta_2}{\Gamma_1!\Gamma_2 \vdash \Delta_1?\Delta_2}\;\text{cut}} \quad\rightsquigarrow\quad \dfrac{\dfrac{\vdots}{\Gamma_1 \vdash A\Delta_1}\quad \dfrac{\vdots}{!\Gamma_2 A \vdash ?\Delta_2}}{\Gamma_1!\Gamma_2 \vdash \Delta_1?\Delta_2}\;\text{cut}$$

(4c)   We have C? in (l) or C! in (r). We have the following situation (note, that in (r) $?A$ was just introduced):

$$\dfrac{\dfrac{\vdots}{\Gamma_1 \vdash ?A?A(?A)^n\Delta_1}\qquad\qquad \vdots}{\dfrac{\Gamma_1 \vdash ?A(?A)^n\Delta_1 \quad !\Gamma_2?A \vdash ?\Delta_2}{\Gamma_1!\Gamma_2 \vdash \Delta_1?\Delta_2}\;\text{cut}} \quad\rightsquigarrow$$

$$\dfrac{\dfrac{\vdots}{\Gamma_1 \vdash ?A?A(?A)^n\Delta_1}\quad \dfrac{\vdots}{!\Gamma_2?A \vdash ?\Delta_2}}{\Gamma_1!\Gamma_2 \vdash \Delta_1?\Delta_2}\;\text{cut}$$

(5)   We combine the proofs Π_1 of (l) and Π_2 of (r) into

$$\star\qquad \Pi_1 = \dfrac{\dfrac{\vdots}{\Gamma_1^0 \vdash A\Delta_1^0}\quad \dfrac{\vdots}{\Gamma_1^1 \vdash B\Delta_1^1}}{\Gamma_1^0\Gamma_1^1 \vdash A \star B\;\Delta_1^0\Delta_1^1} \qquad \Pi_2 = \dfrac{\vdots}{\dfrac{\Gamma_2 AB \vdash \Delta_2}{\Gamma_2\, A \star B \vdash \Delta_2}}$$

$$c(\Pi_1,\Pi_2) = \dfrac{\dfrac{\vdots}{\Gamma_1^1 \vdash B\Delta_1^1}\qquad \dfrac{\dfrac{\vdots}{\Gamma_1^0 \vdash A\Delta_1^0}\quad \dfrac{\vdots}{\Gamma_2 AB \vdash \Delta_2}}{\Gamma_1^0\Gamma_2 B \vdash \Delta_1^0\Delta_2}\;\text{cut}}{\Gamma_1^0\Gamma_1^1\Gamma_2 \vdash \Delta_1^0\Delta_1^1\Delta_2}\;\text{cut}$$

$+$ analogously
$\multimap$ easy
$\sim$ trivial

$$\sqcap \qquad \Pi_1 = \frac{\Gamma_1 \vdash A\Delta_1 \quad \Gamma_1 \vdash B\Delta_1}{\Gamma_1 \vdash A \sqcap B \ \Delta_1} \qquad \Pi_2 = \frac{\Gamma_2 A \vdash \Delta_2}{\Gamma_2 \ A \sqcap B \vdash \Delta_2}$$

$$c(\Pi_1, \Pi_2) = \frac{\Gamma_1 \vdash A\Delta_1 \quad \Gamma_2 A \vdash \Delta_2}{\Gamma_1 \Gamma_2 \vdash \Delta_1 \Delta_2} \ \text{cut}$$

The other case is similar.

$\sqcup$ analogously.

The structure of the proof of strong normalization resembles the argument given for typed lambda calculus. It proceeds by an induction on the complexity of applications of cut, where that complexity is measured principally by the complexity of the cut formula, and further by some measure in ω^2 of the subproofs of the premises of the cut. So the total induction is essentially an induction over ω^3.

21.5. DEFINITION. A *(one step) reduction* of a proof Π is a proof Σ, obtained by applying an appropriate primitive reduction to an instance of the cut rule in Π. Notation $\Pi > \Sigma$ or $\Sigma < \Pi$. $\square$

21.6. LEMMA. *If no reduction applies to a derivation Π then Π is cut free.*

PROOF: As long there is a cut in Π, then it falls in one of the cases listed above; in all those cases a reduction is described, either on the designated cut, or on a related cut (cf. Cases (3c),(3d),(3f)). $\square$

21.7. DEFINITION. Let us define a few notions, in order to get a measure of complexity for cut applications. Define, for Π a derivation terminating in a cut with premises Π_1 and Π_2:

$$a(\Pi_i) = \begin{cases} 0, & \text{if the cut formula is just introduced,} \\ & \quad \text{but not by W?, R?, W!, L!, Ax, R1, L0, R\top, L}\perp \\ 1, & \text{otherwise;} \end{cases}$$

$$a(\Pi) = a(\Pi_1) + a(\Pi_2);$$

$r(\Pi) = $ the number of symbols in the cut formula. $\square$

21.8. DEFINITION. We define the notion of *inductive proof* by induction:

(1) $A \vdash A$; $\vdash \mathbf{1}$; $\mathbf{0} \vdash$; $\Gamma \bot \vdash \Delta$; $\Gamma \vdash \top \Delta$ are inductive proofs;

(2) $\dfrac{\cdots \Gamma_i \vdash \Delta_i \cdots}{\Gamma \vdash \Delta}$ x $\neq$ cut   is inductive if all premises are inductive;

(3) $\Pi = \dfrac{\Pi_1 \quad \Pi_2}{\Gamma \vdash \Delta}$ cut   is inductive if every $\Sigma < \Pi$ is inductive. $\square$

Note that for any Π there are only finitely many $\Sigma < \Pi$. For inductive derivations Π we define the *size* $\mathrm{ind}(\Pi)$ by (the cases match the cases in of the preceding definition.

DEFINITION

(1) $\mathrm{ind}(\Pi) = 1$;
(2) $\mathrm{ind}(\Pi) = \sum_i \mathrm{ind}(\Pi_i) + 1$;
(3) $\mathrm{ind}(\Pi) = \sum_{\Sigma < \Pi} \mathrm{ind}(\Sigma) + 1$. $\square$

21.9. LEMMA. *If Π is inductive, and $\Sigma < \Pi$, then Σ is inductive.*

PROOF: Induction on the structure of Π: if Π is inductive by clause (1) or (2) then Σ is of that form, and the result follows easily from induction hypothesis. If Π is inductive by clause (3) then it follows by the definition of inductive. $\square$

21.10. LEMMA. *Every inductive proof is strongly normalising.*

PROOF: Induction on $\mathrm{ind}(\Pi)$. If $\mathrm{ind}(\Pi) = 1$ then no reductions are possible. If Π is inductive by clause (2) then every reduction is inside one premise. Apply induction hypothesis. If Π is inductive by clause (3) then the result is built into the definition. $\square$

The following lemma is the crucial step towards strong normalisation.

21.11. LEMMA. *If in a proof that ends with a cut, the premises are inductive, then Π is inductive.*

PROOF: Define a complexity of cut applications as follows: given the application with premises $\Gamma_1 \vdash A^n \Delta_1$ and $\Gamma_2 A^m \vdash \Delta_2$ then

$$h(\Pi) := \omega \cdot a(\Pi) + \mathrm{ind}(\Pi_1) + \mathrm{ind}(\Pi_2).$$

First we use induction on $r(\Pi)$, and inside on $h(\Pi)$.

Situation A. Σ arises by reducing a cut in Π_1 or in Π_2. Then we see that $a(\Sigma) \leq a(\Pi)$ and by definition of $\mathrm{ind}(\cdot)$: $\mathrm{ind}(\Sigma) < \mathrm{ind}(\Pi)$. So

$h(\Sigma) < h(\Pi)$; and Σ has inductive premises by Lemma 21.9. Then by h-induction hypothesis, Σ is inductive.

Situation B. Σ arises by reducing the last cut of Π. Then we inspect all possible primitive reductions.

Reduction (1) Σ is one of the premises, and thus inductive by assumption.

Reduction (2) Σ is nearly one of the premises; it is easy to verify that the removal of **1** (resp. **0**) does not affect inductiveness. (See the proof of Lemma 21.2).

Reduction (3a) Σ is inductive by clause (1).

Reduction (3b) The situation is (schematically)

$$\dfrac{\dfrac{\Pi_1'}{\dfrac{\Pi_1 \quad \Pi_2}{\Pi} \text{ cut}}}{} \quad \rightsquigarrow \quad \dfrac{\dfrac{\Pi_1' \quad \Pi_2}{\Pi'} \text{ cut}}{\Sigma}$$

The original cut has complexity

$$h(\Pi) = \omega \cdot (1 + a(\Pi_2)) + \text{ind}(\Pi_1) + \text{ind}(\Pi_2)$$

and the new one

$$h(\Pi') = \omega \cdot (a(\Pi') + a(\Pi_2)) + \text{ind}(\Pi_1') + \text{ind}(\Pi_2)$$

Now $a(\Pi') \leq 1$ and $\text{ind}(\Pi_1') < \text{ind}(\Pi_1)$ so $h(\Pi') < h(\Pi)$. So by h-induction hypothesis Π' is inductive, and then by Definition Σ is inductive.

Reduction (3c) (3e), and (3f) In the same way as in *Reduction (3b)*: One verifies easily that the new cuts have lower h-values then the original one, and concludes that the resulting proof is inductive. For clarity I show *Reduction (3e)*:

$$\dfrac{\dfrac{\Pi_1' \quad \Pi_1''}{\Pi_1} \quad \Pi_2}{\Pi} \text{ cut} \quad \rightsquigarrow \quad \dfrac{\dfrac{\Pi_1' \quad \Pi_2}{\Pi_1^a} \text{ cut} \quad \dfrac{\Pi_1'' \quad \Pi_2}{\Pi_1^b} \text{ cut}}{\Sigma}$$

$$h(\Pi) = \omega \cdot (1 + a(\Pi_2)) + \text{ind}(\Pi_1) + \text{ind}(\Pi_2)$$
$$h(\Pi_1^a) = \omega \cdot (a(\Pi_1') + a(\Pi_2)) + \text{ind}(\Pi_1') + \text{ind}(\Pi_2)$$
$$h(\Pi_1^b) = \omega \cdot (a(\Pi_1'') + a(\Pi_2)) + \text{ind}(\Pi_1'') + \text{ind}(\Pi_2)$$

Now $\text{ind}(\Pi_1'), \text{ind}(\Pi_1'') < \text{ind}(\Pi_1)$ and $a(\Pi_1'), a(\Pi_1'') \leq 1$ so $h(\Pi_1^a)$, $h(\Pi_1^b) < h(\Pi)$. By h-induction hypothesis Π_1^a and Π_1^b are inductive, and by Definition Σ is inductive.

Reduction (4a) Situation (1) is easy, like (3b). Situation (2) is easy, because the cut disappears.

Reduction (4b) Situation (1) We have the following situation:

$$\dfrac{\Pi_1'\quad\Pi_2'}{\dfrac{\Pi_1\quad\Pi_2}{\Pi}\ \text{cut}}\ \rightsquigarrow\ \dfrac{\dfrac{\dfrac{\Pi_1'\quad\Pi_2}{\Pi'}\ \text{cut}\quad\Pi_2'}{\Pi''}\ \text{cut}}{\vdots\atop\Sigma}$$

It is easy to see that the upper cut has lower h-complexity than the original cut; so Π' is inductive. The lower cut has lower rank (its cut formula is A, while the cut formula of the original cut is $?A$), so by r-induction hypothesis Π'' is inductive. So Σ is inductive.

Situation (2): easy.

Reduction (4c) Easy. This is the point where we profit from the extended cut rules!

Reduction (5) Every new cut has lower rank, apply r-induction hypothesis. $\square$

21.12. Corollary. *Every derivation is inductive.*

21.13. Theorem. *(Cut elimination and strong normalisation)*

(1) *The system* **CLL** *is equivalent to the system* **CLL** *without Cut;*

(2) *Every sequence of reductions, applied to a* **CLL**-*proof, terminates.*

Proof: From Corollary 21.12, Lemma 21.10 (2) follows immediately. From this and Lemma 21.6 follows that every proof can be transformed to a cut free proof of the same sequent. $\square$

This theorem is not an adamant result on strong normalisation, because sometimes the application of certain reductions is prohibited (Cases (3c(ii)), (3d), (3f(ii))). There we made the move by excluding those reductions from our set of primitive reductions, thus introducing a bit of context sensitiveness in the concept of primitive reduction. So when we state that strong normalization holds for **CLL**, we should also mention the set of reductions that make this statement true.

References

The following list contains primarily items actually referred to in the text; there is no attempt at a complete bibliography of the rapidly expanding subject. In certain cases I have been unable to obtain full bibliographical data.

ABRAMSKY, S. (1990) Computational interpretations of linear logic, Imperial College Technical Report, DOC 90/20, 52pp. To appear in *Theoretical Computer Science.*

ABRUSCI,V. M. (1990) Sequent calculus for intuitionistic linear propositional logic. In P.P. Petkov, ed., *Mathematical Logic.*, Plenum Press: New York, London.

ASPERTI, A.G.(1990) Categorical Topics in Computer Science. Ph.D. thesis, TD-7/90, Pisa.

AVRON, A. (1988) The semantics and proof theory of linear logic. *Theoretical Computer Science 57*, 161–184.

BENTHEM, J. F. A. K. VAN (1991) *Language in Action: Categories, Lambdas and Dynamic Logic.* North-Holland Publ. Co.: Amsterdam.

BROWN, C. (1990) Linear logic, Petri nets and quantales. Preprint, 16pp.

BROWN, C. AND D. GURR (1990) A categorical linear framework for Petri nets. Preprint, 11pp.

BURNSIDE, W. (1911) *Theory of Groups of Finite Order.* Cambridge University Press: Cambridge, 2nd edition. Reprint Dover Publications, 1955.

CASARI, E. (1987) Comparative logics. *Synthese 73*, 421–449.

CASARI, E. (1989) Comparative logics and abelian l-groups. In R. Ferro, S. Valentini, A. Zanardo, eds., *Logic Colloquium '88*, North-Holland Publ. Co.: Amsterdam, 161–190.

CLOTE, P. (1986) On the finite containment problem for Petri Nets. *Theoretical Computer Science 43*, 99–105.

CURIEN, P.-L. (1986) *Categorical combinators, sequential algorithms and functional programming.* Pitman: London; John Wiley and Sons: New York.

DANOS, V., L. REGNIER (1989) The structure of multiplicatives. *Archive for Mathematical Logic 28*, 181–203.

DANOS, V. (1990) Une application de la logique linéaire à l'étude des processus de normalisation et principalement du lambda calcul. Thèse de Doctorat, Université Paris VII.

DRAGALIN, A. G. (1988) *Mathematical Intuitionism.* American Mathematical Society, Providence RI,USA. Translation of the Russian original from 1979.

DUNN, J. (1986) Relevance logic and entailment. In D. Gabbay, F. Günthner, eds., *Handbook of Philosophical Logic III.* D. Reidel: Dordrecht, 117–224.

ENGBERG, U., G. WINSKEL (1990) Petri nets as models for linear logic. Preprint, 24pp.

FLAGG, R.C., H. FRIEDMAN (1986) Epistemic and intuitionistic formal systems. *Annals of Pure and Applied Logic 32*, 53–60.

GALLIER, J. (1991) Constructive Logics. Part II: Linear Logic and Proof Nets. Research Report 9, Digital Equipment Corporation, Paris.

GIRARD, J.-Y. (1986) The system F of variable types, fifteen years later. *Theoretical Computer Science 45*, 159–192.

GIRARD, J.-Y. (1987) Linear logic. *Theoretical Computer Science 50*, 1–102.

GIRARD, J.-Y. (1988) Multiplicatives. *Rendiconti del Seminario Matematico (Torino)*, Univ. Polit., Torino (special issue on Logic and Computer Science).

GIRARD, J.-Y. (1989) Geometry of interaction 1: Interpretation of system F. In R. Ferro, C. Bonotto, S. Valentini, A. Zanardo, eds., *Logic Colloquium '88*. North-Holland Publ. Co.: Amsterdam, 221–260.

GIRARD, J.-Y. (1990) Geometry of interaction 2: deadlock-free algorithms. In P. Martin-Löf, G.E. Mints, eds., *Colog-88*. Springer Verlag: Berlin, 76–93.

GIRARD, J.-Y. (1991) Quantifiers in linear logic II, Prépublications No. 19, Équipe de Logique Mathématique, Paris VII Logique.

GIRARD, J.-Y., Y. LAFONT, P. TAYLOR (1988) *Proofs and types*. Cambridge University Press: Cambridge, G.B.

GIRARD, J.-Y., A. SCEDROV, P. J. SCOTT (1990) Bounded linear logic: A modular approach to polynomial time computability. In S. Buss, P.J. Scott, eds., *Proceedings of the Mathematical Science Institute Workshop on Feasible Mathematics, Cornell University, June 1988*. Birkhauser Verlag.

GRISHIN, V. N. (1974) A non-standard logic, and its application to set theory. In *Studies in formalized languages and nonclassical logics*. Nauka: Moscow, 135–171.

GRISHIN, V. N. (1981) Predicate and set-theoretic calculi based on logic without contractions. *Math. USSR. Izvestija 18* (translation), 41–59.

HINDLEY, J. R. (1989) BCK-combinators and linear λ-terms have types. *Theoretical Computer Science 64*, 97–105.

HODAS, J. S., D. MILLER (1991) Logic programming in a fragment of intuitionistic linear logic. In *proceedings Sixth Annual IEEE Symposium on Logic in Computer Science, July 15–18,1991, Amsterdam, the Netherlands*. IEEE Computer Society Press, Los Alamitos, California, 32–42.

JAY, C. B. (1990) The structure of free closed categories. *Journal of Pure and Applied Algebra 66*, 271–285.

JUNG, A. (1989) *Cartesian closed categories of domains*. Centre for Mathematics and Computer Science, Amsterdam (CWI Tract 66).

KELLY, G. M. (1964) On MacLane's conditions for coherence of natural associativities, commutativities, etc. *Journal of Algebra 1*, 397–402.

KELLY, G. M., S. MACLANE (1971) Coherence in closed categories. *Journal of Pure and Applied Algebra 1*, 97–140.

KETONEN, J. AND R. WEYHRAUCH (1984) A decidable fragment of predicate calculus. *Theoretical Computer Science 32*, 297–307.

KLEENE, S. C. (1952) *Introduction to Metamathematics.* North-Holland Publ. Co.: Amsterdam.

LAFONT, Y. (1988a) The linear abstract machine. *Theoretical Computer Science 59*, 157–180; Corrections *Ibidem 62* (1988), 327–328.

LAFONT, Y. (1988b) Logiques, Categories et machines. Thèse de doctorat, l' Université Paris VII, 124 pp.

LAFONT, Y. (1988c) Introduction to linear logic. Lecture notes for the Summer School on Constructive Logics and Category Theory, Isle of Thorns, August 1988.

LAFONT, Y., T. STREICHER (1991) Games semantics for linear logic. In *proceedings Sixth Annual IEEE Symposium on Logic in Computer Science, July 15–18,1991, Amsterdam, the Netherlands.* IEEE Computer Society Press, Los Alamitos, California, 43–50.

LAMBEK, J. (1958) The mathematics of sentence structure. *The American Mathematical Monthly 65*, 154–170.

LAMBEK, J. (1961) How to program an infinite abacus. *Canadian Math. Bulletin 4*, 295–302.

LAMBEK, J., P.J. SCOTT (1986) *Introduction to higher-order categorical logic.* Cambridge University Press: Cambridge.

LINCOLN, P. D., J. MITCHELL, A. SCEDROV, N. SHANKAR (1990a) Decision problems for propositional linear logic, Report SRI-CSL-90-08, SRI International Computer Science Laboratory, Menlo Park, California, USA. 76pp.

LINCOLN, P.D., J. MITCHELL, A. SCEDROV, N. SHANKAR (1990b) Decision problems for propositional linear logic. In *Proc. 31st Annual IEEE Symposium on foundations of Computer Science, St. Louis, Missouri, October 1990.* To appear.

LINCOLN, P.D., A. SCEDROV, N. SHANKAR (1991) Linearizing intuitionistic implication. In *proceedings Sixth Annual IEEE Symposium on Logic in Computer Science, July 15–18,1991, Amsterdam, the Netherlands.* IEEE Computer Society Press, Los Alamitos, California, 51–62.

MACLANE, S. (1963) Natural associativity and commutativity. *Rice University Studies 49*, 28–46.

MACLANE, S. (1971) *Categories for the Working Mathematician,* Springer Verlag: Berlin.

MARTÍ-OLIET, N., J. MESEGUER (1989a) An algebraic axiomatization of linear logic models. Report SRI-CSL-89-11, SRI International Computer Science Laboratory, Menlo Park, California, USA. 17pp.

MARTÍ-OLIET, N., J. MESEGUER (1989b) From Petri nets to linear logic. Report SRI-CSL-89-4R2, SRI International Computer Science Laboratory, Menlo Park, California, USA. 37pp.

MARTÍ-OLIET, N., J. MESEGUER (1990) Duality in closed and linear categories. Report SRI-CSL-90-01, SRI International Computer Science Laboratory, Menlo Park, California, USA. 39pp.

MARTÍ-OLIET, N., J. MESEGUER (1991) From Petri nets to linear logic through categories: a survey. To appear.

MAYR, E.W., A. MEYER (1981) The complexity of the finite containment problem for Petri nets. *Journal of the Association for Computing Machinery 28*, 561–576.

MCKINSEY, J. C. C., A. TARSKI (1948) Some theorems about the sentential calculi of Lewis and Heyting. *The Journal of Symbolic Logic 13*, 1–15.

MINSKY, M.L. (1961) Recursive unsolvability of Post's problem of "Tag" and other topics in the theory of Turing machines. *Annals of Mathematics 74*, 437–455.

MINTS, G.E. (1976) Closed categories and the theory of proofs (Russian). *Zapiski Nauchnykh Seminarov LOMI 68*, 83–114. Translated *Journal of Soviet Mathematics 15* (1981), 45–62.

OEHRLE, R. T., E. BACH, D. WHEELER , eds. (1988) *Categorial grammars and Natural Language Structures.* D. Reidel Publ. Co.: Dordrecht.

ONO, H. (1989) Algebraic aspects of logics without structural rules. To appear in the Proceedings of the Malcev Conference, Novosibirsk, August 1989.

ONO, H. (1990a) Structural rules and a logical hierarchy. In P.P. Petkov, ed., *Mathematical Logic.* Plenum Press: New York, London, 95–104.

ONO, H. (1990b), Phase structures and quantales — a semantical study of logics without structural rules. To appear (lecture delivered at the conference "Logics with restricted logical rules", University of Tübingen, October 1990).

ONO, H., Y. KOMORI (1985) Logics without the contraction rule. *The Journal of Symbolic Logic 50*, 169–201.

PAIVA, V.C.V. DE (1988), The Dialectica Categories. Ph.D. thesis, University of Cambridge. Also, slightly modified, as Report 213, Computer Laboratory, University of Cambridge (1991).

PAIVA, V.C.V. DE (1989), The Dialectica categories. In J.W. Gray and A. Scedrov, eds., *Categories in computer Science and Logic.* American Mathematical Society: Providence, RI. (Also in *Contemporary Mathematics* 92, 47–62).

ROORDA, D. (1989) Investigations in classical linear logic. ITLI-prepublication series ML-89-03, University of Amsterdam, 19pp.

ROORDA, D. (1990) Quantum graphs and proof nets. Manuscript, June 1990.

ROORDA, D. (1991) Resource logics: proof-theoretical investigations. Academisch Proefschrift (Ph.D. thesis), Universiteit van Amsterdam.

SAMBIN, G. (1989) Intuitionistic formal spaces and their neighbourhoods. In R. Ferro, C. Bonotto, S. Valentini, A. Zanardo, eds., *Logic Colloquium '88.* North-Holland Publ. Co.: Amsterdam, 221–245.

SCOTT, D.S. (1982) Domains for denotational semantics. In M. Nielsen, E. M. Schmidt, eds., *Automata, Languages and Programming. 9th Colloquium.* Springer Verlag: Berlin (Springer Lecture Notes in Computer Science 140), 577–613.

SEELY, R.A.G. (1989) Linear logic, *-autonomous categories, and cofree coalgebras. *Contemporary Mathematics 92,* 371–382.

SCHELLINX, H. (1990) Some syntactical observations on linear logic. ITLI-prepublication series. Report ML-90-08, University of Amsterdam, 25pp. To appear in *Logic and Computation.*

TROELSTRA, A.S. , ed. (1973) *Metamathematical Investigation of Intuitionistic Arithmetic and Analysis.* Springer Verlag: Berlin.

YETTER, D. (1990) Quantales and (non-commutative) linear logic. *The Journal of Symbolic Logic 55,* 41–64.

Index

197

CSLI Publications

Lecture Notes

The titles in this series are distributed by the University of Chicago Press and may be purchased in academic or university bookstores or ordered directly from the distributor at 5501 Ellis Avenue, Chicago, Illinois 60637.

A Manual of Intensional Logic. Johan van Benthem, second edition, revised and expanded. Lecture Notes No. 1. ISBN 0-937073-29-6 (paper), 0-937073-30-X (cloth)

Emotion and Focus. Helen Fay Nissenbaum. Lecture Notes No. 2. ISBN 0-937073-20-2 (paper)

Lectures on Contemporary Syntactic Theories. Peter Sells. Lecture Notes No. 3. ISBN 0-937073-14-8 (paper), 0-937073-13-X (cloth)

An Introduction to Unification-Based Approaches to Grammar. Stuart M. Shieber. Lecture Notes No. 4. ISBN 0-937073-00-8 (paper), 0-937073-01-6 (cloth)

The Semantics of Destructive Lisp. Ian A. Mason. Lecture Notes No. 5. ISBN 0-937073-06-7 (paper), 0-937073-05-9 (cloth)

An Essay on Facts. Ken Olson. Lecture Notes No. 6. ISBN 0-937073-08-3 (paper), 0-937073-05-9 (cloth)

Logics of Time and Computation. Robert Goldblatt. Lecture Notes No. 7. ISBN 0-937073-12-1 (paper), 0-937073-11-3 (cloth)

Word Order and Constituent Structure in German. Hans Uszkoreit. Lecture Notes No. 8. ISBN 0-937073-10=5 (paper), 0-937073-09-1 (cloth)

Color and Color Perception: A Study in Anthropocentric Realism. David Russel Hilbert. Lecture Notes No. 9. ISBN 0-937073-16-4 (paper), 0-937073-15-6 (cloth)

Prolog and Natural-Language Analysis. Fernando C. N. Pereira and Stuart M. Shieber. Lecture Notes No. 10. ISBN 0-937073-18-0 (paper), 0-937073-17-2 (cloth)

Working Papers in Grammatical Theory and Discourse Structure: Interactions of Morphology, Syntax, and Discourse. M. Iida, S. Wechsler, and D. Zec (Eds.) with an Introduction by Joan Bresnan. Lecture Notes No. 11. ISBN 0-937073-04-0 (paper), 0-937073-25-3 (cloth)

Natural Language Processing in the 1980s: A Bibliography. Gerald Gazdar, Alex Franz, Karen Osborne, and Roger Evans. Lecture Notes No. 12. ISBN 0-937073-28-8 (paper), 0-937073-26-1 (cloth)

Information-Based Syntax and Semantics. Carl Pollard and Ivan Sag. Lecture Notes No. 13. ISBN 0-937073-24-5 (paper), 0-937073-23-7 (cloth)

Non-Well-Founded Sets. Peter Aczel. Lecture Notes No. 14. ISBN 0-937073-22-9 (paper), 0-937073-21-0 (cloth)

Partiality, Truth and Persistence. Tore Langholm. Lecture Notes No. 15. ISBN 0-937073-34-2 (paper), 0-937073-35-0 (cloth)

Attribute-Value Logic and the Theory of Grammar. Mark Johnson. Lecture Notes No. 16. ISBN 0-937073-36-9 (paper), 0-937073-37-7 (cloth)

The Situation in Logic. Jon Barwise. Lecture Notes No. 17. ISBN 0-937073-32-6 (paper), 0-937073-33-4 (cloth)

The Linguistics of Punctuation. Geoff Nunberg. Lecture Notes No. 18. ISBN 0-937073-46-6 (paper), 0-937073-47-4 (cloth)

Anaphora and Quantification in Situation Semantics. Jean Mark Gawron and Stanley Peters. Lecture Notes No. 19. ISBN 0-937073-48-4 (paper), 0-937073-49-0 (cloth)

Propositional Attitudes: The Role of Content in Logic, Language, and Mind. C. Anthony Anderson and Joseph Owens. Lecture Notes No. 20. ISBN 0-937073-50-4 (paper), 0-937073-51-2 (cloth)

Literature and Cognition. Jerry R. Hobbs. Lecture Notes No. 21. ISBN 0-937073-52-0 (paper), 0-937073-53-9 (cloth)

Situation Theory and Its Applications, Vol. 1. Robin Cooper, Kuniaki Mukai, and John Perry (Eds.). Lecture Notes No. 22. ISBN 0-937073-54-7 (paper), 0-937073-55-5 (cloth)

The Language of First-Order Logic (including the Macintosh program, Tarski's World). Jon Barwise and John Etchemendy, second edition, revised and expanded. Lecture Notes No. 23. ISBN 0-937073-74-1 (paper)

Tarski's World. Jon Barwise and John Etchemendy. Lecture Notes No. 25. ISBN 0-937073-67-9 (paper)

Situation Theory and Its Applications, Vol. 2. Jon Barwise, J. Mark Gawron, Gordon Plotkin, Syun Tutiya, editors. Lecture Notes No. 26. ISBN 0-937073-70-9 (paper), 0-937073-71-7 (cloth)

Literate Programming. Donald E. Knuth. Lecture Notes No. 27. ISBN 0-937073-80-6 (paper), 0-937073-81-4 (cloth)

Other CSLI Titles Distributed by UCP

Agreement in Natural Language: Approaches, Theories, Descriptions. Michael Barlow and Charles A. Ferguson (Eds.). ISBN 0-937073-02-4 (cloth)

Papers from the Second International Workshop on Japanese Syntax. William J. Poser (Ed.). ISBN 0-937073-38-5 (paper), 0-937073-39-3 (cloth)

The Proceedings of the Seventh West Coast Conference on Formal Linguistics (WCCFL 7). . ISBN 0-937073-40-7 (paper)

The Proceedings of the Eighth West Coast Conference on Formal Linguistics (WCCFL 8). . ISBN 0-937073-45-8 (paper)

The Phonology-Syntax Connection. Sharon Inkelas and Draga Zec (Eds.) (co-published with The University of Chicago Press). ISBN 0-226-38100-5 (paper), 0-226-38101-3 (cloth)

The Proceedings of the Ninth West Coast Conference on Formal Linguistics (WCCFL 9). . ISBN 0-937073-64-4 (paper)

Japanese/Korean Linguistics. Hajime Hoji (Ed.). ISBN 0-937073-57-1 (paper), 0-937073-56-3 (cloth)

Experiencer Subjects in South Asian Languages. Manindra K. Verma and K. P. Mohanan (Eds.). ISBN 0-937073-60-1 (paper), 0-937073-61-X (cloth)

Grammatical Relations: A Cross-Theoretical Perspective. Katarzyna Dziwirek, Patrick Farrell, Errapel Mejías Bikandi (Eds.). ISBN 0-937073-63-6 (paper), 0-937073-62-8 (cloth)

The Proceedings of the Tenth West Coast Conference on Formal Linguistics (WCCFL 10). . ISBN 0-937073-79-2 (paper)

Books Distributed by CSLI

The Proceedings of the Third West Coast Conference on Formal Linguistics (WCCFL 3). . ($10.95) ISBN 0-937073-45-8 (paper)

The Proceedings of the Fourth West Coast Conference on Formal Linguistics (WCCFL 4). . ($11.95) ISBN 0-937073-45-8 (paper)

The Proceedings of the Fifth West Coast Conference on Formal Linguistics (WCCFL 5). . ($10.95) ISBN 0-937073-45-8 (paper)

The Proceedings of the Sixth West Coast Conference on Formal Linguistics (WCCFL 6). . ($13.95) ISBN 0-937073-45-8 (paper)

Hausar Yau Da Kullum: Intermediate and Advanced Lessons in Hausa Language and Culture. William R. Leben, Ahmadu Bello Zaria, Shekarau B. Maikafi, and Lawan Danladi Yalwa. (*$19.95*) ISBN 0-937073-68-7 (paper)

Hausar Yau Da Kullum Workbook. William R. Leben, Ahmadu Bello Zaria, Shekarau B. Maikafi, and Lawan Danladi Yalwa. (*$7.50*) ISBN 0-93703-69-5 (paper)

Ordering Titles Distributed by CSLI

Titles distributed by CSLI may be ordered directly from CSLI Publications, Ventura Hall, Stanford University, Stanford, California 94305-4115 or by phone (415)723-1712 or (415)723-1839. Orders can also be placed by e-mail (pubs@csli.stanford.edu) or FAX (415)723-0758.

All orders must be prepaid by check, VISA, or MasterCard (include card name, number, expiration date). For shipping and handling add $2.50 for first book and $0.75 for each additional book; $1.75 for the first report and $0.25 for each additional report. California residents add 7% sales tax.

For overseas shipping, add $4.50 for first book and $2.25 for each additional book; $2.25 for first report and $0.75 for each additional report. All payments must be made in US currency.

CSLI Lecture Notes report new developments in the study of language, information, and computation. In addition to lecture notes, the series includes monographs and conference proceedings. Our aim is to make new results, ideas, and approaches available as quickly as possible.